Generis
PUBLISHING

AF837239

Smart Construction Safety

Artificial Intelligence for Risk Prevention and Site Management

Dr. Rubén Rodríguez Elizalde

Title: **Smart Construction Safety**

Artificial Intelligence for Risk Prevention and Site Management

ISBN: 979-8-89966-027-6

Author: Dr. Rubén Rodríguez Elizalde

Cover image: www.pixabay.com

Publisher: Generis Publishing
Online orders: www.generis-publishing.com
Contact email: info@generis-publishing.com

Smart Construction Safety
Artificial Intelligence for Risk Prevention and Site Management

TABLE OF CONTENTS

Prologue

We are living in an era where technology advances faster than our ability to assimilate it. Artificial intelligence has ceased to be a science fiction concept and has become a tangible reality—present in our homes, our smartphones, and increasingly, in our workplaces. Yet, amid discussions of productivity, efficiency, and automation, there is one dimension that cannot—and must not—be overlooked: the safety of people.

Occupational risk prevention has always been a cornerstone of dignified and responsible work. Thanks to it, millions of lives have been safeguarded, and many work environments have become safer. But the challenges have not disappeared. On the contrary: new hazards are emerging, along with new forms of work organization and novel technologies which, if not properly managed, can create more problems than they solve.

This book is born out of a clear conviction: artificial intelligence can be a powerful ally in prevention, if used with prudence, ethics, and purpose. It is not about replacing humans, but about enhancing their ability to see further, anticipate more accurately, and act with greater precision. It is, ultimately, a tool in the service of life.

From the foundational concepts of artificial intelligence to its specific applications in the construction industry—one of the sectors most exposed to occupational accidents—this work aims to provide a comprehensive and practical perspective for any professional seeking to take the next step in occupational health and safety management.

Those of us who have worked in the field of prevention know that small details can be decisive in saving lives. This book invites us to look beyond: to embrace innovation as a core component of our preventive culture and to build a future where technology and people do not compete for space, but rather coexist in a complementary way, serving the collective well-being.

Introduction

Justification for the Book: Why It Is Necessary

Artificial Intelligence (AI) is one of the most disruptive technologies driving contemporary digital transformation. Its capacity to automate cognitive processes, optimize complex systems, and enable new models of interaction has accelerated its cross-sector adoption in strategic areas such as healthcare, finance, manufacturing, and logistics.

This advancement has triggered a profound reconfiguration of the labor and organizational landscape, creating both unprecedented opportunities and structural challenges. One of the most pressing and widely debated issues in this context concerns the net effect AI will have on employment—in other words, answering the question: Will artificial intelligence displace more jobs than it creates?

Despite its progress, the integration of artificial intelligence in fields such as occupational risk prevention remains in an early stage of technological maturity. Its implementation is still relatively limited and often misinterpreted—both operationally and from a regulatory standpoint. This not only represents an implementation gap but also a strategic opportunity to modernize preventive practices through predictive, adaptive, and evidence-based computational approaches.

In a labor market where technological, organizational, and social changes unfold rapidly, occupational risk prevention (ORP) must evolve to keep pace with these shifts. It is no longer sufficient to simply comply with regulations or conduct periodic inspections. It is necessary to anticipate risks, personalize preventive measures, and make decisions based on real-time, data-driven insights.

This book was conceived to address a gap: to provide a rigorous, practical, and human-centered perspective on how artificial intelligence can be embedded into an organization's preventive culture. More importantly, it shows how AI can save lives and prevent harm when applied with critical thinking and clear intent.

Current State of Occupational Risk Prevention

In many organizations, preventive management continues to face persistent challenges:

- Reactive approach: action is taken only after an accident occurs.

- Lack of structured and analyzed data.
- Rigid and bureaucratic procedures.
- Difficulty in supervising complex or rapidly changing environments (such as construction sites).
- Limited personalization in training and applied safety measures.

Despite regulatory and technological progress, occupational accident rates remain high in many sectors. Traditional tools, on their own, are insufficient to address modern risks, which now include psychosocial factors, multitasking, partial automation, an aging workforce, and extreme environmental conditions.

Brief Introduction to Artificial Intelligence (AI)

Artificial Intelligence is a branch of computer science that develops systems capable of performing tasks that typically require human intelligence—such as learning, pattern recognition, reasoning, prediction, communication, and decision-making.

Through techniques such as machine learning, deep learning, and computer vision, AI enables the processing of large volumes of data, the detection of risks invisible to the human eye, and the automation of preventive processes with high precision.

In the workplace context, this translates into:

- Real-time monitoring.
- Predictive risk assessment.
- Automation of safety audits.
- Monitoring of personal protective equipment (PPE) usage.
- Detection of fatigue, distraction, or hazardous behavior.

Objectives of the Book

The main objective of this book is to explain how artificial intelligence can be ethically and effectively integrated into occupational risk prevention, and how such integration can transform not only outcomes but also the overall workplace safety culture.

Specific Objectives:

1. Present the technical foundations of artificial intelligence in an accessible manner.

2. Analyze the current framework and challenges of occupational risk prevention.
3. Explore real-world and emerging applications of artificial intelligence in occupational health and safety.
4. Identify benefits, limitations, and associated risks.
5. Examine in depth the specific case of the construction sector.
6. Propose a roadmap for responsible and effective implementation.

This book is intended for safety professionals, engineers, innovation managers, educators, students, and anyone interested in the intersection of technology, safety, and workplace well-being.

Methodological Framework

The development of this book is based on an interdisciplinary and applied methodological approach, aimed at combining conceptual rigor with practical utility. The main objective has been to offer a clear, up-to-date, and operational guide on artificial intelligence applied to occupational risk prevention, both for university students and active professionals seeking to understand and adopt these technologies.

Approach and Sources

The methodology adopted is supported by three complementary pillars:

1. Specialized literature review: A systematic analysis was conducted of academic, technical, and regulatory literature related to artificial intelligence and occupational risk prevention. Sources include scientific articles, reference books, reports from official bodies (INSST, EU-OSHA, NIOSH), current legislation, and sector-specific technical documents. This documentary foundation has enabled the construction of a solid and up-to-date framework for developing the content.
2. Case study analysis: Six relevant case studies of AI implementation in workplace environments—particularly in the construction sector—were selected and analyzed. These cases were chosen for their exemplary potential, demonstrated technical feasibility, and instructional value. They include solutions focused on predictive analytics, computer vision, fatigue detection, and audit automation.

3. Consultation with industry professionals: Informal exchanges were maintained throughout the book's development with safety technicians, engineers, innovation leads, and specialized trainers. These contributions helped identify real-world barriers, unmet needs, and key perceptions regarding the level of technological adoption in the field of prevention.

Structure and Criteria

The content has been organized with a progressive logic that begins with conceptual and regulatory foundations (Chapters 1 and 2), moves toward practical applications (Chapter 3), examines benefits and limitations (Chapters 4 and 5), and culminates in a sector-specific focus (Chapter 6), real-world examples (Chapter 7), and future trends (Chapter 8). This structure is designed to facilitate gradual understanding by combining theoretical foundations with operational implications.

For the selection of technologies and case studies, the following criteria were applied:

- A sufficient technology readiness level (TRL $\geq$ 6).
- Accessible and transparent documentation.
- Potential for direct application in real workplace settings.
- Demonstrated or promising relevance for prevention.

Limitations of the Approach

This book does not intend to replace regulatory manuals or provide a one-size-fits-all formula for implementing artificial intelligence in occupational risk prevention. Its limitations include:

- The absence of proprietary quantitative data or longitudinal impact studies.
- A focus on industrial sectors, which may limit direct applicability to other fields.
- The potential obsolescence of certain technological developments due to the rapid pace of innovation.

Despite these limitations, the fundamental purpose of this work is to offer a critical and guiding foundation that enables professionals and students to approach artificial intelligence from a realistic, responsible perspective—one that aligns with the core values of prevention: anticipation, protection, and continuous improvement.

1 Foundations of Artificial Intelligence

1.1 What Is Artificial Intelligence?

Artificial Intelligence (AI) is an advanced branch of computer science dedicated to developing systems capable of performing tasks that, until recently, could only be carried out by humans (Alastruey, 2021). These include capabilities such as reasoning, learning, planning, adapting to change, interpreting visual or auditory input, and even understanding and generating natural language.

However, AI is not an isolated discipline. It exists at the intersection of numerous fields of knowledge, including mathematics, statistics, neuroscience, control theory, logic, computational linguistics, and even philosophy (Ramírez Véliz et al., 2022). This interdisciplinary foundation enables the creation of what are known as artificial agents: systems capable of perceiving their environment, making decisions, and acting accordingly—even in complex or dynamic situations (Ríos, 2023).

To better understand how it works, AI can be divided into several specializations, each with its own methodological focus:

- Machine Learning (*ML*): This refers to a machine's ability to learn from data (Alcobilla Ferrara, 2023). In other words, machines do not need to be programmed step by step for each task; instead, they learn from experience and improve over time.
- Natural Language Processing (*NLP*): This allows systems to understand, generate, and communicate using human language (Celi-Parraga et al., 2021). Thanks to this technology, we have increasingly accurate chatbots, automatic translators, and virtual assistants.
- Computer Vision: This field enables machines to "see" and interpret images and video. It allows systems to recognize objects, movements, and situations—an especially useful capability in surveillance, industrial quality control (Lopez C. et al., 2018), and occupational safety.
- Intelligent Robotics: This refers to the application of AI in physical machines (Mendizá & Jiménez, 2021). These are robots capable of perceiving their surroundings, making autonomous decisions, and acting based on environmental input.

Together, these areas constitute a technological ecosystem that extends the capabilities of machines beyond traditional limits. Today, artificial intelligence is present in a wide array of sectors—from industry and healthcare to education, mobility, and, of course, occupational risk prevention (Macías García, 2022).

1.1.1 Formal Conceptualization

From a functional perspective, artificial intelligence can be understood as a set of methods, algorithms, and computational structures designed to enable machines to behave intelligently (Alastruey, 2021). We refer to systems—or *artificial agents*—that are capable of perceiving what is happening around them (via sensors), processing that information, and acting upon the environment (via actuators), all in an autonomous and adaptive manner; that is, by learning and adjusting according to circumstances.

One of the most widely recognized frameworks for classifying different approaches to artificial intelligence is that proposed by researchers Stuart Russell and Peter Norvig, two leading figures in the academic field (Bohannon, 2015). According to them, artificial intelligence can be understood through four major system types (Russell, 2017) (Fig. 1):

1. Systems that think like humans: These aim to replicate mental processes, such as reasoning or decision-making.
2. Systems that act like humans: These aim to reproduce human behavior, though not necessarily the cognitive processes behind it.
3. Systems that think rationally: These follow logical models to reach correct or coherent conclusions.
4. Systems that act rationally: These make optimal decisions based on available data and defined objectives, without necessarily resembling human behavior.

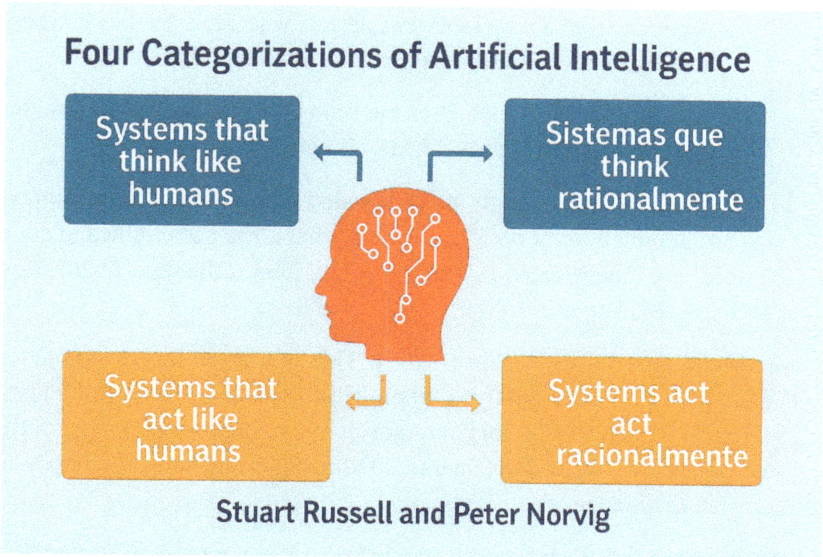

Fig. 1. Infographic illustrating AI approaches according to Russell and Norvig (graphic by the author).

In practice, most applications currently found in industrial or professional environments fall into this last category (Caisa Herrera et al., 2024): systems that act rationally. In other words, technologies that analyze data, identify patterns, and optimize decision-making with the goal of improving processes, reducing errors, or increasing safety.

1.1.2 Main Categories of AI

Artificial Intelligence can be classified into three major levels based on its ability to learn and generalize beyond specific tasks:

- Weak or Narrow AI: This is the most common and widely used form of AI today (Llata et al., 2000; López de Mántaras, 2020). It refers to systems designed to perform a specific task, such as object detection, data classification, or generating a specific prediction. These systems lack consciousness and a comprehensive understanding of their environment (Ríos, 2023); they operate within well-defined boundaries.

  Typical examples include:

- A computer vision system that detects whether a worker is wearing a safety helmet (Rodríguez Elizalde, 2025).
- An algorithm that estimates the probability of workplace accidents based on historical data (Macías García, 2022).

This type of AI cannot apply its knowledge beyond the specific context it was programmed for. It performs well in what it has been trained to do, but it does not understand anything beyond that (Niebles, 2020; Reyes Cárdenas, 2023).

- Artificial General Intelligence (*AGI*): This is a more theoretical concept referring to AI with cognitive capabilities comparable to those of a human being. An AGI would be able to reason across multiple contexts, learn from new situations (López de Mántaras, 2020), understand language deeply, and transfer knowledge across domains.

  As of today, Artificial General Intelligence does not exist (Carrera, 2024). It remains a long-term research goal with significant interest, but we are still far from achieving it technically. Some experts even question whether it will ever be technically feasible (Carabantes López, 2013).

- Artificial Superintelligence (*ASI*): This refers to a hypothetical and currently speculative scenario (Mir, 2023). ASI would surpass human cognitive capabilities in all aspects—from creativity and ethical reasoning (Antúnez Cid, 2024) to strategic thinking and complex decision-making.

  Although this type of AI is not a reality, it raises important ethical and societal questions (Otero Parga, 2023; Rodríguez Elizalde, 2025; Terrones Rodríguez, 2018):

  - Who would control it?
  - How would it be aligned with human interests?
  - What implications would it have for society and employment?

1.1.3 Capacidades operativas de los sistemas inteligentes

Current artificial intelligence systems do not function as "magic boxes" that perform everything at once. They are designed to simulate complex cognitive functions (such as perceiving, reasoning, learning, or decision-making), but achieve this by breaking these functions into specific capabilities, each supported by mathematical, statistical, and computational models.

These capabilities can be grouped into several domains which, taken together, form what we might call the cognitive cycle of artificial intelligence (Alastruey, 2021; Castro Castro et al., 2022) (Fig. 2):

- Perception: capturing what occurs in the environment.
- Knowledge representation: organizing perceived information.
- Reasoning: drawing conclusions based on structured data.
- Decision-making: selecting the optimal option after evaluating multiple alternatives according to an optimization criterion.
- Machine learning: improving performance through experience.
- Natural interaction: communicating effectively with humans.

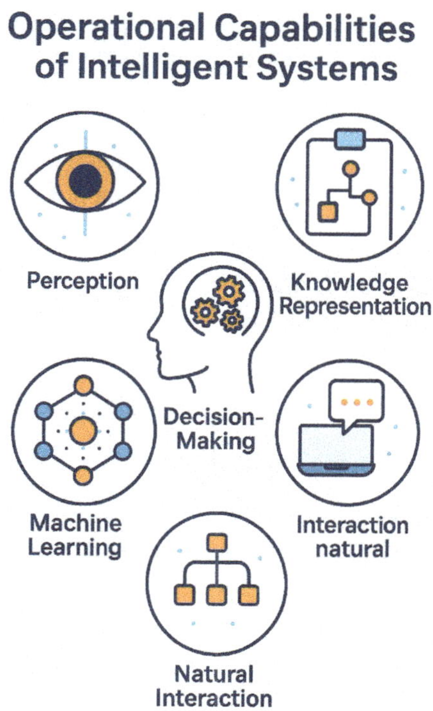

Fig. 2. Infographic illustrating the operational capabilities of intelligent systems (graphic by the author).

These capabilities are interrelated and are made possible by fundamental advances in algorithms, hardware architectures (such as GPUs and TPUs), and the massive availability of data (Big Data).

Below, each of these capabilities is explored individually:

1. Perception: Capturing what is happening in the environment. This is the system's input gateway. Artificial intelligence can "perceive" through sensors (Freré Arauz et al., 2022; Ramírez Véliz et al., 2022) that collect visual, auditory, thermal, or environmental data. This data is then interpreted accordingly:
    - Computer vision: Processing and analyzing images or real-time video. Enables detection of personal protective equipment (PPE), recognition of warning signs, or identification of hazardous movements.
    - Auditory perception: Identification of relevant sounds, such as alarms, anomalous noises, or human voices.
    - Environmental sensors: Measure temperature, gases, vibrations, etc.

    This stage enables the system to build an initial representation of the environment, which is crucial for subsequent processes.
2. Knowledge Representation: Organizing what has been perceived (Wagman, 2021). Once it has data, the system must interpret it. To do so, it organizes the information into internal structures: environmental maps, cause-effect relationships, or work contexts. For example:
    - Recognizing that a zone is active because people are working in it.
    - Linking fatigue to task overload or extended shifts.
3. Reasoning: Drawing conclusions (Ortega Giménez et al., 2021). Here, the system combines what it has perceived with prior knowledge to generate new inferences. It can do this through various methods:
    - Deductive reasoning: From general to specific, typical of expert systems.
    - Inductive reasoning: Learning from observed data, typical of machine learning.
    - Probabilistic reasoning: Calculating likelihoods, highly useful in uncertain or dynamic environments.

- In occupational safety, this might mean detecting an imminent risk in a specific area by combining multiple real-time variables.

4. Decision-Making: Choosing the best option. Once the situation is analyzed, the system must act—determining what to do, how, and when (Chicaiza Ortiz, 2024). It does so by combining:

   - Its objectives—such as minimizing risk, optimizing time, and reducing cost.
   - Its understanding of the environment.

   The potential outcomes of each action:

   - Rule-based decision systems (or expert systems). These are computational structures that make decisions based on a predefined set of logical rules. They operate under the classical "if this condition occurs, then perform that action" scheme (IF-THEN rules).

     In practice, these systems codify the knowledge and experience of human experts into a structured rule base. An inference engine then evaluates the current environmental conditions and applies the corresponding rules to determine the most appropriate course of action.

     These systems are especially effective in well-defined, structured environments where logic is clear and predictable—such as in industrial processes or standardized procedures.

     However, they present limitations: in environments marked by uncertainty, ambiguity, or low structure—as is often the case in occupational risk prevention—these systems may prove too rigid or insufficiently flexible.

   - Bayesian networks. These are graphical models designed to handle uncertainty in a structured manner. They represent a set of variables and their dependency relationships using a Directed Acyclic Graph (*DAG*). In such graphs:

     - Each node represents a variable (e.g., "noise level" or "accident risk").
     - Each edge indicates a directional influence between variables.

- Each node is associated with a conditional probability table (*CPT*) that computes the likelihood of a variable's state given the states of its parent nodes.

These networks allow for probabilistic inference—that is, estimating the likelihood of an event under conditions of partial or uncertain information. Probabilities can be updated dynamically as new data become available, following Bayes' theorem (Tuya Gonzalez et al., 2011).

Bayesian networks are particularly useful in decision-making contexts where certainty is not attainable—such as in occupational risk scenarios.

- Multi-objective optimization algorithms. In many real-world situations, it is insufficient to optimize a single objective (Bastidas et al., 2010). For instance, one might aim to reduce costs without compromising safety, or to improve efficiency without undermining worker well-being.

  Multi-objective optimization addresses this by seeking a balance among potentially conflicting objectives (Boligán Rojas et al., 2024). Rather than a single best solution, it identifies a set of optimal trade-offs—known as the Pareto front—where improving one objective necessarily entails a compromise in another.

  Algorithms commonly used for this purpose include NSGA-II, MOEA/D, and SPEA2, which operate on populations of candidate solutions and apply evolutionary or heuristic methods to explore the solution space.

  These tools are widely applied in fields such as engineering, logistics, economics, and increasingly, in occupational risk management.

- Reinforcement Learning (*RL*). This type of learning is inspired by human trial-and-error behavior. An agent (i.e., an intelligent system) interacts with its environment, makes decisions, and receives feedback in the form of rewards or penalties based on the outcomes of its actions (Montenegro Meza et al., 2023).

  Unlike supervised learning, RL does not rely on labeled data to indicate correct outcomes. Instead, the agent learns through experience—by acting, observing results, and iteratively refining its behavior.

Formally, this is modeled using a Markov Decision Process (*MDP*). The goal is to learn an optimal policy (π^*) that prescribes the best possible action in each state to maximize cumulative long-term reward (Portillo-Ramírez et al., 2024; Van Otterlo, 2009).

Reinforcement learning is especially effective in dynamic, uncertain environments with delayed feedback—such as robotics, gaming, or complex adaptive control systems.

5. Machine Learning: A cornerstone of AI is its learning capability. Rather than repeating preset routines, the system adjusts its behavior based on outcomes. Learning types include (Retto-Manrique, 2021):

    - Supervised learning: Trains on labeled examples (e.g., "risk" / "no risk").
    - Unsupervised learning: Discovers hidden patterns without labels
    - Semi-supervised and self-supervised learning: Use incomplete or sparsely labeled data.
    - Reinforcement learning: Learns through action and feedback in dynamic environments.

    These methods make AI increasingly accurate and adaptable over time.

6. Natural Interaction: Communicating with humans. Modern systems don't just analyze data—they can also interact naturally (Celi-Parraga et al., 2021), enhancing workplace integration. Key features include:

    - Natural Language Processing (*NLP*): Understanding and generating human language.
    - Multimodal interfaces: Combine text, images, gestures, or voice.
    - Virtual assistants: Answer questions, perform evaluations, and support decision-making in real time.

    These interaction capabilities significantly improve usability and user acceptance.

These capabilities do not operate in isolation. They function continuously as part of a natural cycle: perceiving, understanding, reasoning, deciding, learning, and interacting. This is how AI becomes a powerful ally—especially in complex fields like occupational risk prevention.

From environmental monitoring to real-time decision-making, AI is capable of redefining how we understand and manage workplace risk (Goñi Sein, 2024; Macías García, 2022, 2023; Nájar Becerra, 2024). In short, AI does not replace traditional prevention—it complements and enhances it, helping us anticipate threats more effectively, act more judiciously, and, ultimately, protect the most valuable asset of any organization: its people.

1.1.4 Artificial Intelligence and Traditional Automation

Artificial intelligence and automation are often mistakenly treated as interchangeable concepts. However, although both technologies aim to improve processes and enhance task efficiency through computational systems, they operate in fundamentally different ways and offer distinct levels of adaptability and application.

The distinction between the two is not merely technical; understanding it clearly is essential to selecting the appropriate solution in each context—especially in complex work environments such as occupational risk prevention (Ficapal Cusí, 2024; Goñi Sein, 2024).

1.1.4.1 Traditional Automation: Deterministic Logic and Structural Rigidity

Traditional automation is based on executing a predefined set of instructions in a sequential and explicit manner. In other words, the system performs actions following a deterministic and predictable logic: each input corresponds to a specific output—"if A happens, do B" (Ponce Solé, 2024).

This model—typically implemented through conditional logic, flowcharts, or supervisory control and data acquisition (SCADA) systems—is highly effective for repetitive and standardized tasks that operate within known and stable parameters (Pérez-López, 2015). Classic examples include the automated control of a production line or the triggering of an alarm when a threshold value is exceeded.

However, its main limitation is clear and significant: it cannot adapt to the unexpected. When faced with a situation not foreseen in its original programming, the system lacks the capacity for autonomous response (López de Mántaras, 2020;

Rojas, 2024). In such cases, it may cease functioning properly or require immediate human intervention to regain control.

1.1.4.2 Artificial Intelligence: Flexibility, Learning, and Complex Decision-Making

Unlike traditional automation, artificial intelligence is not limited to following fixed rules. What makes it distinctive is its incorporation of learning mechanisms, knowledge representation, and autonomous inference capabilities (Alastruey, 2021). These features allow it to operate effectively even in changing, uncertain, or poorly structured environments.

An artificial intelligence system is capable of analyzing large volumes of data, detecting patterns, learning from experience, adjusting its internal models, and improving its decisions continuously (De Paula, 2023; Diestra Quinto et al., 2021) (Table 1).

Table 1. Comparative summary: Artificial Intelligence vs. Traditional Automation (table by the author)

Feature	Traditional Automation	Artificial Intelligence
Operational model	Sequential, rule-based	Adaptive, data- and learning-based
Applicable environments	Stable and predictable	Dynamic, high variability
Adaptability	Very limited	High (via continuous feedback)
Uncertainty management	None or marginal	Intrinsic (through probabilistic models)
Human intervention required	Frequent in novel situations	Minimal—system learns and self-adjusts
Example in OSH	Alarm triggered by threshold	Accident prediction via complex pattern recognition

For example, while an automated system may simply shut down a machine when a temperature threshold is exceeded, an AI-based system can anticipate a thermal failure before it occurs—analyzing real-time variables such as vibration, humidity, workload, or usage history.

Moreover, it can autonomously adjust parameters to prevent the issue from materializing. This capacity for anticipation and adaptation makes artificial intelligence an essential tool for decision-making in complex and ever-evolving contexts (Zapata Cortés, 2020).

Traditional automation is highly effective for well-defined tasks in stable environments, but it falls short in the face of novel or dynamic conditions. Artificial intelligence, on the other hand, is designed to manage complexity, learn from its surroundings, and make autonomous, real-time decisions.

Therefore, in today's work environments—characterized by uncertainty, variability, and the constant need for rapid response (Blasco Jover, 2024)—artificial intelligence emerges as a key enabler of future-ready occupational safety and health systems (Macías García, 2022).

1.1.5 AI in the Workplace Context

Artificial intelligence is not an isolated technology; as detailed in section 1.1.1 (Formal Conceptualization), AI is in fact a complex network of computational methods, algorithms, and application environments that enable systems to process information, learn from data, and perform advanced cognitive tasks autonomously and adaptively. Due to its ability to operate in changing, uncertain, and highly complex environments, AI surpasses many limitations of traditional automation by incorporating decision-making models based on statistical inference, deep learning, and continuous improvement (Antezana Soria Galvarro, 2023).

This versatility has positioned AI as a key driver of digital transformation across diverse sectors such as healthcare, finance, industry, transportation, agriculture, logistics, e-commerce, and education. Its impact extends beyond efficiency, redefining processes, improving decision-making, reducing information asymmetries, and enabling new forms of personalization, prediction, and computational autonomy. However, this rapid advancement is not without significant challenges that require critical reflection and a balanced approach between innovation and responsibility.

One of the main obstacles faced by AI is its high dependency on data. Contemporary systems, particularly those based on deep learning, require massive, structured, and representative datasets to function correctly. This need increases development costs and introduces a significant risk: if data are incomplete, mislabeled, or biased, the system may learn inaccurately, producing erroneous, unfair, or even dangerous outcomes (Schor-Landman, 2022; Torres Cadillo, 2023).

Compounding this issue is the high computational demand. Training complex models involves intensive calculation processes that can only be executed with advanced infrastructures such as GPUs or TPUs. This not only limits access for many small and medium-sized enterprises but also raises key questions about sustainability: the high energy consumption associated with AI development and deployment necessitates a reconsideration of how to balance innovation with environmental efficiency (Terrones Rodríguez, 2018).

Another critical limitation is the capacity for generalization. Many models perform well with the data they were trained on but exhibit unpredictable behavior in new situations. This phenomenon, known as overfitting, hinders practical AI applications in real-world contexts where conditions frequently change. Moreover, knowledge transfer between tasks remains limited, often requiring models to be retrained nearly from scratch for each new scenario.

A further concern is the lack of interpretability (Reyes Cárdenas, 2023). As models become more powerful, they also become more opaque. In particular, deep learning systems often function as black boxes: they produce effective outcomes, but without clear explanations of how those outcomes were derived. This lack of transparency can undermine user trust, complicate validation in regulated environments, and hinder adoption in sectors where decision traceability is crucial, such as healthcare, law, or occupational safety.

Beyond technical aspects, AI raises significant ethical and social issues. One of the most pressing is algorithmic bias (Chicaiza Ortiz, 2024; Lassi, 2023; Schor-Landman, 2022). Because systems learn from historical data, if those data reflect societal or cultural inequalities, the model is likely to replicate and even amplify them. This issue has already been documented in applications such as hiring, credit scoring, and judicial risk assessment. Achieving real algorithmic equity requires more than data cleansing—it demands a fundamental reevaluation of training criteria and frameworks.

Closely related is the concern over privacy (Rodríguez-Martínez et al., 2023). AI applications that involve personal data require stringent information management. Compliance with data protection laws is a starting point, but not sufficient: systems must incorporate privacy-by-design principles and adopt techniques that ensure data is handled securely, ethically, and transparently.

There are also security risks associated with AI systems themselves. AI can be vulnerable to external manipulation, such as adversarial attacks, where input data are subtly altered to cause system errors. This poses a serious threat, especially in systems operating in critical environments like autonomous vehicles or automated industrial infrastructures.

Accountability is another crucial issue. As AI systems make decisions with direct impacts on individuals, it is essential to clearly define who is responsible for the outcomes. Without proper assignment of responsibility among developers, operators, and users, there is a risk of creating legal voids that erode trust in the technology and hinder effective governance.

Last but not least is the socio-economic impact of AI. Its ability to automate tasks—particularly repetitive or routine ones—may lead to significant job displacement. Addressing this challenge requires decisive political action: accessible technology training, workforce reskilling programs, and redistributive models to ensure that the benefits of innovation reach everyone, not just dominant corporate or geopolitical actors.

The evolution of artificial intelligence is as rapid as it is promising. But its development must be accompanied by deep reflection on its technical, ethical, and social implications. It is not merely about creating more efficient or accurate systems but ensuring that their impact is fair, inclusive, and sustainable. For AI to truly serve human progress, it must be designed, implemented, and regulated according to principles of responsibility, equity, and transparency.

Looking to the future with ambition does not mean ignoring risks. On the contrary, it involves recognizing the tremendous opportunities AI offers in sectors such as healthcare, education, justice, and occupational risk prevention—and embracing them with the conviction that every technological advance embodies a vision of society (Terrones Rodríguez, 2018). The key lies in building an innovation ecosystem that not only drives development but does so while respecting individual rights, reinforcing collective trust, and placing technology at the service of the common good.

1.2 Brief History of Artificial Intelligence

The history of artificial intelligence is, fundamentally, the history of how we have learned to understand, model, and replicate processes inherent to the human mind through machines (León Cuenca et al., 2024). From its roots in classical philosophy to its consolidation as a scientific discipline, AI has followed a path shaped by major theoretical advances, technological leaps, and alternating cycles of enthusiasm and disillusionment. Understanding this evolution not only clarifies the current capabilities and limitations of AI, but also helps anticipate its future role (León García, 2023) in critical domains such as occupational risk prevention—where its applications can fundamentally reshape how we anticipate and protect.

Although the idea of understanding and reproducing intelligent behavior has deep philosophical roots (from Aristotle's formal logic to the mechanistic ideals of the Renaissance), modern artificial intelligence is grounded in mathematical and computational advances from the 20th century (Ekmekci & Arda, 2020). A key turning point came in 1936, when Alan Turing proposed the concept of a Universal Machine—a theoretical model that laid the groundwork for modern computing and, for the first time, allowed us to conceive of machines capable of executing any computable function (Leguizamón Páez & Becerra Correa, 2024).

Just a few years later, in 1950, Turing advanced further with his seminal paper Computing Machinery and Intelligence, in which he posed the then-radical question: Can machines think? To explore this, he proposed what would become the famous Turing Test—a method to evaluate whether a machine could imitate human behavior to the point of being indistinguishable in conversation. That question and the methodological approach it introduced marked the formal starting point of what we now understand as artificial intelligence (Ekmekci & Arda, 2020; Leguizamón Páez & Becerra Correa, 2024; Pino V. et al., 2022).

The institutional and disciplinary foundation of AI was laid in 1956, when John McCarthy, Marvin Minsky, Claude Shannon, and Nathaniel Rochester convened the now-legendary Dartmouth Conference. It was at this event that the term "Artificial Intelligence" was first coined, along with the foundational hypothesis that "every aspect of learning or any other feature of intelligence can in principle be so precisely described that a machine can be made to simulate it" (Ekmekci & Arda, 2020). This marked the birth of AI as an autonomous academic field.

In the years that followed, research focused mainly on what is now known as symbolic AI. Early programs like the Logic Theorist and General Problem Solver demonstrated basic logical reasoning capabilities (Ekmekci & Arda, 2020; Leguizamón Páez & Becerra Correa, 2024; Pino V. et al., 2022). Expert systems also emerged, encoding human knowledge into explicit rule-based structures—typically represented as IF-THEN statements.

Despite initial success, these systems had clear limitations. They were effective in well-defined environments but struggled to adapt to novel situations. Their scalability was poor, and they lacked robust mechanisms for handling uncertainty. Nevertheless, they laid the conceptual and technical foundations for future innovations.

The initial wave of AI enthusiasm soon gave way to periods of stagnation known as the AI winters. The first, from 1974 to 1980, stemmed from the inability of symbolic systems to cope with real-world complexity and ambiguity (Leguizamón Páez & Becerra Correa, 2024; Pino V. et al., 2022). The second, from 1987 to 1993, was driven by the commercial decline of expert systems—despite early promise, they proved costly to maintain, difficult to scale, and inadequate in dynamic environments.

Starting in the 1990s, AI experienced a renaissance powered by advances in statistical techniques and the rise of machine learning. Researchers began to move away from symbolic reasoning toward data-driven models. Algorithms such as Support Vector Machines (SVMs), decision trees, and ensemble methods like Random Forest and AdaBoost proved effective in classification and prediction tasks, achieving unprecedented accuracy (Leguizamón Páez & Becerra Correa, 2024; Pino V. et al., 2022).

This methodological shift was reinforced by two key trends: the explosion of digital data—thanks to the internet, sensors, and interconnected systems—and continuous improvements in computing power through faster, more efficient processors. These advances enabled the training of larger, more complex models and ushered in practical applications such as search engines, recommendation systems, and AI-assisted medical diagnostics.

A major turning point came in 2012, when the deep convolutional neural network AlexNet won the ImageNet competition with overwhelming superiority in large-scale image classification. This marked the dawn of the deep learning era and brought renewed focus to neural networks, which had been largely sidelined due to computational constraints.

From that point on, deep learning not only redefined the state of the art in computer vision, speech recognition, and natural language processing—it also paved the way for increasingly powerful and versatile models. Technologies like transformers enabled systems such as GPT, BERT, AlphaGo, and ChatGPT to generate coherent text (Gutierrez-Vasques & Mijangos de la Cruz, 2024), translate languages, master complex games, or create images from text prompts (Alastruey, 2021). Just a few years prior, such capabilities belonged to the realm of science fiction.

This qualitative leap has not been driven by algorithmic advances alone. Key enablers include cloud-based distributed infrastructures, specialized hardware like GPUs and TPUs, and the rise of an open-source culture that democratized access to tools, data, and models (López Eguiarte et al., 2024). As a result, both large research labs and independent communities have been able to experiment and deploy AI solutions at unprecedented speed.

Today, AI research advances along multiple frontiers: enhancing model interpretability (XAI), increasing robustness to adversarial inputs, minimizing environmental impact, improving knowledge transfer across tasks, and ensuring ethical data governance. Beyond these technical aims, fundamental issues are gaining traction—such as how to prevent algorithmic bias, the role of human oversight, accountability in automated decisions, and the global governance of large-scale AI systems.

What is clear is that AI has transitioned from a niche discipline to a core component of today's digital infrastructure. Its history—summarized in Table 2—is marked by cycles of innovation, stagnation, and consolidation. And it offers a key lesson: technical progress alone is not enough. For artificial intelligence to truly drive human progress, it must be guided by critical reflection, institutional commitment, and a long-term vision that balances innovation with responsibility.

Table 2. Chronological Development of Key Milestones in AI History (compiled by the author).

Period	Key Milestone	Summary Description
Classical Philosophy – 19th c.	Formal logic and mechanistic thinking	Early ideas about logical reasoning and simulating intelligence with mechanisms.
1936	Turing Machine	Alan Turing introduces the theoretical model of universal computation, foundational to modern computing.
1950	*Computing Machinery and Intelligence*	Turing asks whether machines can think and proposes the Turing Test as an operational benchmark.
1956	Dartmouth Conference	The term *Artificial Intelligence* is coined and AI is formally established as an academic discipline.
1956–1970	Symbolic AI and early expert systems	Development of logic-based programs and rule-based systems simulating expert reasoning.
1974–1980	First AI Winter	Disappointment with symbolic models' failure in real-world problems.
1987–1993	Second AI Winter	Commercial collapse of expert systems, decline in funding and interest.
1990–2010	Revival through machine learning	Emergence of statistical algorithms, data explosion, and computational growth.
2012	AlexNet's victory in ImageNet	Start of the deep learning era with convolutional neural networks in computer vision.
2017–present	Rise of transformers and generative models	Emergence of GPT, BERT, and DALL·E with advanced language, visual, and reasoning capabilities.
Present	Consolidation and ethical challenges	AI as critical infrastructure; focus on explainability, sustainability, fairness, privacy, and governance.

1.3 Relevant Types of AI Algorithms

Algorithms constitute the true operational core of any artificial intelligence system (González Guerrero, 2021; Landa Arroyo, 2021). They are the formal structures that enable data processing, pattern recognition, modeling of complex relationships, and autonomous decision-making. Everything an intelligent system does—from recognizing an image to generating text or predicting a technical failure—ultimately depends on a well-designed algorithm.

Artificial intelligence is not a single technique, but rather a broad and diverse set of interconnected methodologies that allow machines to learn from experience, adapt to new contexts, and operate with a certain degree of autonomy. From a functional standpoint, these methodologies are grouped into different branches, among which machine learning has become the central pillar of modern artificial intelligence.

Among the main techniques, the following stand out (Boligán Rojas et al., 2024; González Guerrero, 2021; Lassi, 2023; Lopez C. et al., 2018; Schor-Landman, 2022):

- Machine learning, in which the system learns from data without the need to predefine all the rules. This can include:
    - Supervised learning, where the model is trained on labeled data (e.g., images classified as "hazardous" or "non-hazardous").
    - Unsupervised learning, which detects patterns without prior labels—useful for exploratory analysis or data segmentation.
    - Reinforcement learning, which learns through trial and error by optimizing decisions based on accumulated rewards.
- Deep learning, a more advanced form of machine learning based on deep neural networks. It is particularly effective in tasks involving large volumes of unstructured data, such as images, text, or audio.
- Natural Language Processing (NLP), which enables machines to understand, generate, and interact using human language. It has practical applications in training, virtual assistance, document analysis, and internal workplace communication.
- Computer vision, which allows systems to interpret visual content (images, video, real-time sequences). It is a key tool for automated site supervision, visual risk detection, and monitoring the proper use of personal protective equipment (PPE).

The following sections will elaborate on the main families of algorithms that underpin contemporary artificial intelligence, with particular emphasis on their technical foundations, representative applications, and the limitations that still condition their deployment in real-world environments.

1.3.1 Machine Learning

Machine learning is one of the most prominent and impactful areas within artificial intelligence. Its core principle is the development of algorithms capable of automatically learning from data, without the need to explicitly program all the rules that govern behavior. Instead of instructing the machine on how to act in every situation, data is provided and the system is trained to detect patterns and make decisions on its own (L. D. Jones et al., 2018; Soori et al., 2023).

This learning process can take different forms depending on the type of data available and the specific objective being pursued (Pineda, 2022; Soori et al., 2023):

- Supervised learning involves training the system on labeled data, i.e., examples where the correct outcome is known in advance. For instance, the model may be shown images of construction sites classified as "fall risk" or "no risk," with each case properly labeled. The algorithm then adjusts its internal parameters to minimize prediction error. This is the most widely used approach for classification, regression, and automated diagnosis tasks.
- Unsupervised learning is applied when labeled data are not available. Here, the system's goal is to autonomously discover hidden structures within the data—such as clusters, internal relationships, or latent patterns that are not immediately obvious. This method is especially useful for exploratory data analysis, segmentation, or dimensionality reduction, particularly when predefined categories do not exist or the task does not require a definitive answer.
- Reinforcement learning, by contrast, operates under a different paradigm. In this case, the agent does not learn from labeled examples but rather through interaction with a dynamic environment. It learns via trial and error, receiving rewards or penalties based on the outcomes of its actions. The goal is to optimize behavior over time by maximizing the cumulative reward. This approach is particularly well-suited for complex, non-deterministic environments such as robotics, adaptive logistics, or video games.

1.3.2 Deep Learning

Deep learning is an advanced branch of machine learning that leverages artificial neural networks with multiple layers to model highly complex and nonlinear relationships. These networks are composed of hierarchical processing structures that enable the system to extract and abstract deep-level features from data—even when such data are unstructured (Alastruey, 2021).

Thanks to this architecture, deep learning is particularly effective in tasks involving images, audio, or language, where meaningful patterns are not readily apparent or are difficult to encode through manual programming.

In the field of occupational risk prevention, deep learning is becoming increasingly relevant across a variety of use cases. Some of the most notable applications include (Alastruey, 2021; Sugil, 2024):

- Real-time automated image classification to monitor work environments such as construction sites or industrial zones.
- Facial recognition and detection of proper use of personal protective equipment (PPE), enabling regulatory compliance monitoring without the need for constant human oversight.
- Automatic speech-to-text transcription and multilingual translation, especially useful in multilingual settings or when verbal instructions need to be documented in real time.

The power of deep learning lies in its ability to learn directly from raw data, without the need to manually define each rule, and to adapt to complex or changing situations with a high degree of accuracy.

1.3.3 Natural Language Processing (NLP)

Natural Language Processing (NLP) is a branch of artificial intelligence focused on enabling interaction between machines and human language, whether in written or spoken form (Sancho Escrivá et al., 2020). To achieve this, NLP systems rely on statistical, semantic, and syntactic models that allow them to interpret, generate, or classify messages in a coherent and context-aware manner (Gutierrez-Vasques & Mijangos de la Cruz, 2024).

In workplace environments, NLP opens the door to highly practical applications in occupational risk prevention, both for daily operations and for enhancing internal communication. Some prominent examples include (Celi-Parraga et al., 2021; Sancho Escrivá et al., 2020):

- Virtual assistants specialized in safety, capable of answering queries regarding regulations, safety protocols, or incident response procedures.
- Automated generation of risk assessment reports, streamlining administrative tasks and minimizing human error.
- Sentiment analysis of workplace climate surveys, helping to detect emotional states, stress, or discomfort within teams in order to anticipate organizational or psychosocial issues.

These systems are typically built on supervised learning techniques, deep neural networks, and advanced architectures such as transformers—among which models like GPT and BERT stand out for their ability to understand context and generate language with a remarkable degree of naturalness (Gutierrez-Vasques & Mijangos de la Cruz, 2024; López Eguiarte et al., 2024).

Thanks to NLP, artificial intelligence not only understands what we say, but can also communicate effectively and with discernment, adapting to the technical and human language that characterizes professional environments.

1.3.4 Computer Vision

Computer vision enables machines to interpret visual information from images, video footage, or real-time sequences. This is made possible through advanced deep learning algorithms, particularly convolutional neural networks (CNNs), which are specifically designed to identify objects, people, actions, and contextual situations in real-world environments with high accuracy (Laad et al., 2024).

In the field of occupational risk prevention, this technology offers a broad range of highly valuable applications that significantly enhance safety and responsiveness in dynamic or high-risk work environments. Among the most notable applications are the following (Anagnostopoulos & Krinidis, 2025; Massiris et al., 2021):

- Automated surveillance of construction sites, enabling the real-time detection of unauthorized access or anomalous situations.
- Verification of the correct use of personal protective equipment (PPE) through video-based image analysis, facilitating compliance monitoring without the need for constant human oversight.
- Recognition of hazardous postures, abrupt movements, or imminent falls, allowing preventive action to be taken before an accident occurs.

Thanks to these capabilities, computer vision has become a key tool for anticipating risks, reducing exposure to hazards, and fostering safer and more controlled working environments.

Table 3. Conceptual Summary: AI vs. Traditional Automation (table by the author)

Functional Category	Technical Type	Key PRL Applications
Machine Learning	Supervised	Risk situation classification
	Unsupervised	Anomaly detection, exploratory analysis
	Reinforcement	Adaptive control of inspection robots/drones
Deep Learning	Neural networks	Visual hazard detection, speech recognition
NLP	Language models	Safety assistants, automated document analysis
Computer Vision	CNNs and segmentation	Real-time visual monitoring, PPE compliance control

The various AI algorithms and technologies should not be viewed as competing approaches, but as complementary tools chosen based on task nature, data availability, and specific objectives.

In occupational risk prevention, this combination enables smart, integrated solutions capable of perception, contextual analysis, and automated decision-making within a single system. This synergy between models, techniques, and applications is what gives AI its true transformative power in complex and dynamic work environments.

1.4 The Potential of AI in Industry

Artificial intelligence has established itself as a first-tier enabling technology in the digital transformation of industry. Its ability to analyze large volumes of data, detect complex patterns, automate cognitive processes, and adapt to changing environments positions it as a key strategic element for optimizing operations, improving efficiency, and anticipating failures in critical systems.

Unlike other supporting digital technologies—such as automation software, SCADA systems, or the Internet of Things—artificial intelligence introduces a layer of decision-making and adaptability that allows it to go beyond the mere automatic execution of tasks (Chernyshev et al., 2022; Pérez-López, 2015; Stefana et al., 2023). AI-based systems do not simply follow predefined instructions: they learn, predict, recommend, and act based on context and accumulated experience.

Thanks to these capabilities, artificial intelligence facilitates a paradigm shift: from a reactive to a proactive approach, where intervention occurs before the incident, not after. This shift does not only affect isolated tasks; it redefines the operational logic of production systems, steering them towards more adaptive, predictive, and autonomous models.

For organizations that manage to integrate artificial intelligence with strategic vision, ethical commitment, and a sense of responsibility, this technology will not be just another tool, but a true engine of sustainable competitive advantage. In the field of occupational risk prevention, that value can translate directly into lives saved, injuries avoided, and safer working environments for everyone.

1.4.1 Operational Optimization and Predictive Efficiency

One of the areas where the impact of artificial intelligence is most tangible is in the optimization of industrial processes, whether in production environments or advanced logistics systems (Boligán Rojas et al., 2024; Chen Cheng et al., 2023). Through the use of supervised and unsupervised learning algorithms, industrial plants can now incorporate analytical and adaptive capabilities that, until recently, required constant human supervision.

Key applications include (Caisa Herrera et al., 2024):

- Real-time bottleneck detection, enabling dynamic resource reallocation and task rescheduling.
- Demand and energy consumption forecasting, allowing for more efficient and sustainable planning.
- Automated adjustment of production parameters without human intervention, improving precision and reducing error margins.
- Reduction of downtime through predictive maintenance techniques, anticipating failures before they disrupt operations.

This approach facilitates the transition towards the "zero unplanned downtime" paradigm by leveraging models capable of predicting mechanical, electrical, or structural failures in advance (Mohan et al., 2023). These systems analyze weak signals captured by sensors—such as abnormal vibrations, thermal oscillations, or slight performance deviations—that would go unnoticed in conventional systems.

The convergence of artificial intelligence and advanced sensor technology is transforming how industrial systems are designed, operated, and maintained, ushering in a new generation of processes that are more efficient, resilient, and autonomous (Anagnostopoulos & Krinidis, 2025).

1.4.2 Automated Quality and Intelligent Control

In traditional quality control systems, inspections are often based on manual visual reviews or physical sampling, which can introduce variability and limitations in terms of speed and consistency. Artificial intelligence allows these methods to be replaced or supplemented with technologies such as computer vision, automated classification, and anomaly detection in finished products (Chen Cheng et al., 2023; Schor-Landman, 2022).

These solutions not only accelerate the inspection process but also provide tangible improvements in precision, traceability, and consistency, even under variable conditions of lighting, shape, color, or texture. Algorithms trained on real examples can detect minute defects or atypical patterns that might escape the human eye, contributing to reduced rework, scrap, and associated costs (Tito Chura, 2023).

Moreover, AI models can autonomously adjust production parameters based on quality outcomes, thus closing the feedback loop and enabling more fine-tuned and adaptive control of the system. This approach not only enhances the final product but also turns quality into a dynamic component of the process itself, rather than a static check at the end of the production line (Alastruey, 2021).

1.4.3 Cyber-Physical Systems and Adaptive Autonomy

The integration of artificial intelligence into cyber-physical systems (CPS) and collaborative robotics environments is giving rise to increasingly autonomous and decentralized industrial models. In these new scenarios, robots, autonomous vehicles, robotic arms, and internal transport systems can coordinate their actions through reinforcement learning algorithms or multi-agent planning, allowing for localized and adaptive decision-making (Peña-Cáceres et al., 2025).

This autonomy does not imply the elimination of the human factor. Rather, it represents an evolution in the role of the operator—from direct task execution to strategic supervision, scenario validation, workflow design, and algorithmic quality control. Ultimately, this is a model of human–machine collaboration where artificial intelligence does not replace, but instead augments the operational capabilities of the entire system (Cogliati et al., 2018; Tito Chura, 2023).

1.4.4 Risk Modelling and Intelligent Industrial Safety

In the field of occupational risk prevention, artificial intelligence offers transformative potential across two complementary dimensions.

On the one hand, from a preventive standpoint, AI enables the anticipation of incident likelihood by analyzing historical, environmental, organizational, and human factors. This predictive capability supports dynamic planning of work shifts, task allocation, and risk zone delimitation, adapting in real time to the actual conditions of each operational context.

On the other hand, from a reactive perspective, AI-based systems can issue real-time alerts through technologies such as computer vision, smart sensors, or automated contextual analysis. This allows for the detection of unsafe behaviors, regulatory non-compliance, or hazardous conditions—such as heat stress, unsafe postures, or incorrect use of personal protective equipment—before they result in harm to workers.

This approach paves the way toward adaptive safety, where the protection of personnel no longer relies solely on adherence to static regulations, but evolves according to the changing context of the workplace, combining anticipation, flexibility, and responsive capacity.

1.4.5 Strategic Potential and Data-Driven Decision-Making

At the corporate level, artificial intelligence enables a shift from decision-making based on intuition or past experience to a fully data-driven approach. Strategic decisions are grounded in predictive models, trend analysis, and simulations of future scenarios. This shift not only enhances operational efficiency but also reduces uncertainty and strengthens organizational resilience in the face of increasingly volatile and dynamic environments.

In this context, the use of prescriptive models—those that not only predict what might happen but also recommend the best course of action—will become increasingly prominent. Their application will be especially critical in areas such as supply chain management, advanced production planning, or dynamic resource allocation, where the ability to anticipate and respond optimally can create significant competitive advantages.

1.4.6 Summary of Industrial Impact

The potential of artificial intelligence in the industrial sector can be synthesized around five key strategic vectors, reflecting its capacity to comprehensively transform both productive and organizational processes:

- Productivity, by maximizing operational performance through more efficient use of resources, minimizing downtime, and optimizing every phase of the process.
- Flexibility, by enabling systems to respond quickly and autonomously to changing conditions—whether due to demand fluctuations, supply chain disruptions, or internal incidents.
- Quality, by improving consistency in outcomes, enhancing product traceability, and enabling early detection of defects or deviations—even in early stages of production.
- Safety, by anticipating and mitigating risks from both occupational health and operational continuity perspectives, thereby reducing exposure to accidents (Macías García, 2022) or critical failures.
- Contextual intelligence, by making it possible to interpret complex data in real time, resulting in faster, more informed, and operationally aligned decision-making.

These vectors, summarized in Table 4, do not operate in isolation but rather reinforce each other synergistically, making artificial intelligence a cornerstone technology for driving a more efficient, resilient, and future-oriented industry.

Table 4. Strategic Vectors of AI Impact in Industry (table by the author).

Strategic Vector	AI Contribution
Productivity	Optimizes operational performance and resource utilization, reducing inefficiencies.
Flexibility	Enables fast, autonomous adaptation to changes in the environment or demand.
Quality	Enhances traceability, enables early error detection, and strengthens consistency.
Safety	Anticipates and mitigates occupational and operational risks through predictive analysis and sensors.
Contextual Intelligence	Interprets complex data in real time for more accurate and agile decision-making.

1.5 Current Limitations

Despite the remarkable technical advances and the growing presence of artificial intelligence across a wide range of sectors, its current state remains far from complete or universally robust. While its practical applications have multiplied over the past two decades, there are still significant technical, operational, and structural constraints that must be acknowledged to avoid premature, biased, or unsustainable deployments (Carrera, 2024).

These limitations do not negate its potential, but they do define the realistic scope within which AI can operate effectively and responsibly. This is particularly crucial in sensitive domains—such as occupational risk prevention, which is the focus of this book—where understanding these limitations is key to designing safe, reliable, and context-appropriate systems.

One of the most critical barriers is the heavy reliance on data. Contemporary models, especially those based on deep learning, require massive amounts of labeled data to function adequately. This presents several concurrent challenges (Alawamleh et al., 2024; Walton, 2018):

- Data collection, cleaning, and annotation are resource-intensive tasks that demand significant investment.
- Incomplete, imbalanced, or biased data negatively impact model performance and introduce hard-to-detect distortions.

- In critical sectors such as healthcare, law, or occupational safety, data is often scarce, sensitive, or heterogeneous, making its use even more complex.

Furthermore, a structural limitation persists: models do not truly understand the meaning of data—they only learn statistical correlations. This means they may produce seemingly valid outputs without genuinely comprehending why they are correct, leaving them vulnerable to contextual shifts or unforeseen situations.

Compounding this are issues of computational limitations and scalability. Training advanced models requires specialized infrastructure—such as GPUs, TPUs, or distributed computing systems (Alawamleh et al., 2024; Walton, 2018). This entails significant barriers to entry:

- Small and medium-sized enterprises often lack access to such resources.
- Training large models consumes considerable energy, raising concerns over sustainability and carbon footprint.
- Additionally, bigger models do not necessarily yield better results—scaling remains a technical challenge with often diminishing returns.

Although cloud-based solutions have lowered the threshold for accessing high-performance computing, the ability to train and deploy cutting-edge models remains concentrated in the hands of a few key players, creating significant global imbalances.

Another fundamental limitation is the restricted generalization capability. Many models perform well within the specific contexts for which they were trained but fail when exposed to new or different data—a problem known as overfitting. Moreover, transferring learning across tasks remains an unresolved issue. Techniques like transfer learning or few-shot learning have made strides, but most systems are still narrow-task focused, requiring retraining under new conditions.

A further concern is the lack of explainability and transparency, especially in powerful deep neural network systems. These models often behave like true black boxes: they deliver highly accurate results, but their internal reasoning is opaque. This lack of interpretability presents several issues (Alawamleh et al., 2024; Walton, 2018):

- It undermines user trust, particularly in high-stakes domains involving lives or rights.
- It complicates regulatory audits and validation, which are essential for responsible adoption.

- It hampers legal traceability when accountability must be established in the event of error or harm.

While the field of Explainable AI (XAI) is advancing, it still lacks widely adopted, stable solutions suitable for complex operational environments.

Lastly, current models exhibit weak reasoning and contextual rigidity. Their function is rooted in correlation, not causal logic or deep semantic understanding. As a result:

- They fail to infer general rules beyond their training data.
- They struggle with ambiguity, novel scenarios, or dynamic environments.
- They have difficulty integrating and reconciling multiple sources of information coherently.

This makes human supervision indispensable in tasks requiring common sense, ethical reasoning, or complex human interaction.

These limitations can be grouped into five major categories, summarized in Table 5.

Table 5. Main Structural Limitations of Contemporary AI (table by the author).

Axis	Description
Data dependency	Requires large volumes of high-quality labeled data. Problems arise with incomplete, biased, or sensitive datasets. Learning is based on correlation, not causality.
Computational constraints	Needs specialized hardware (GPU, TPU), high energy consumption, and access to advanced infrastructures. Entry barriers for SMEs and technological concentration.
Limited generalization	Poor adaptability to new contexts or tasks. High risk of overfitting. Transfer learning remains immature and expensive.
Lack of explainability	Complex models are difficult to interpret ("black boxes"). Issues with trust, regulatory compliance, and legal accountability.
Weak reasoning and contextual rigidity	Lacks semantic and causal understanding. Struggles with ambiguity, integrating information, or operating with common sense.

In summary, despite the spectacular progress of recent years, artificial intelligence continues to face structural technical limitations that constrain its reliability, applicability, and scalability. Recognizing and understanding these constraints is not a theoretical exercise, but a prerequisite for designing realistic, safe, and ethically responsible systems (Carrera, 2024).

This is especially vital in safety-critical environments, such as occupational risk prevention, where values like trust, transparency, and contextual adaptation are not optional—they are essential. Only with this critical lens—blending innovation with caution—can we build AI solutions that genuinely add value and align with the foundational principles of workplace safety and human well-being.

2 Occupational Risk Prevention (ORP)

2.1 What Is Occupational Risk Prevention?

Occupational Risk Prevention (ORP) is the set of strategies, disciplines, systems, and practices designed to ensure the protection of workers' health, safety, and well-being by identifying, assessing, controlling, and mitigating hazards associated with professional activities. It is not merely a regulatory obligation, nor a reactive response to accidents; it is a multidimensional discipline—technical, legal, organizational, and ethical—that seeks to anticipate risk and foster a proactive safety culture across all levels of the workplace.

Occupational risk prevention (ORP) is a multidisciplinary field—encompassing technical, legal, and organizational dimensions—focused on proactively safeguarding workers' health, safety, and overall well-being in the workplace In its modern conception, prevention goes beyond avoiding immediate accidents or physical injuries: it involves a comprehensive management approach to all risk factors—physical, chemical, biological, ergonomic, organizational, and psychosocial—that may impair an employee's functional capacity in the short, medium, or long term (Monereo Pérez & López Insua, 2016).

Technically, occupational risk prevention consists of a set of actions, procedures, and methodologies applied across all stages of production or service delivery. Its purpose is to identify, assess, eliminate, or control occupational hazards systematically. This is rooted in the core principle that workplace safety is not a static condition, but a dynamic process of continuous improvement.

From a regulatory standpoint, most national frameworks—regardless of region—are increasingly aligned with international principles, such as those established by the International Labour Organization (ILO), the World Health Organization (WHO), and various ISO standards (e.g., ISO 45001). These frameworks emphasize prevention as a fundamental right of all workers and establish minimum standards for risk assessment, health surveillance, training, and participation.

At its core, ORP is grounded in a holistic view of occupational health, which extends far beyond the prevention of physical injuries. It addresses a wide range of risk factors that can compromise an individual's functional capacity or mental and physical health over time. These include:

- Physical risks, such as noise, vibrations, radiation, or mechanical hazards.

- Chemical and biological risks, including exposure to toxic substances, pathogens, or allergens.
- Ergonomic risks, such as repetitive strain, poor posture, or inadequate equipment.
- Psychosocial risks, including stress, burnout, violence, or work-related mental health issues.
- Organizational risks, like shift mismanagement, poor communication, or unclear responsibilities.

This comprehensive scope reflects the reality that workplace hazards are increasingly complex, interrelated, and dynamic, especially in environments shaped by globalization, digitalization, and the integration of emerging technologies like artificial intelligence, robotics, and cyber-physical systems.

2.1.1 A Science and a Strategy: The Foundations of ORP

From a technical and methodological standpoint, ORP entails the continuous application of structured processes across the entire life cycle of work—from planning and design, to execution, monitoring, and evaluation. These processes include:

- Hazard identification: Recognizing anything with the potential to cause harm.
- Risk assessment: Estimating the likelihood and severity of harm.
- Risk control: Implementing preventive and protective measures to reduce risk to acceptable levels.
- Monitoring and review: Continuously evaluating and adjusting the effectiveness of controls.

These activities are carried out using evidence-based methods, quantitative and qualitative tools, and sector-specific standards and regulations.

From a legal and regulatory perspective, most national occupational safety systems are aligned—either directly or indirectly—with international conventions and frameworks, including:

- The International Labour Organization (ILO) conventions, particularly Convention No. 155 (Occupational Safety and Health, 1981) and

Convention No. 187 (Promotional Framework for Occupational Safety and Health, 2006).

- The World Health Organization (WHO) Global Plan of Action on Workers' Health.
- The UN Sustainable Development Goals (SDG 8.8), which call for "safe and secure working environments for all workers."
- The ISO 45001:2018 international standard for Occupational Health and Safety Management Systems (OHSMS), which provides a global benchmark for systematic and continuous improvement in worker safety.

Many countries incorporate these frameworks into national legislation, while supranational entities (e.g., the European Union) provide legally binding directives—such as the EU Framework Directive 89/391/EEC—requiring member states to ensure minimum preventive standards.

2.1.2 From Hazard to Culture: A Systemic Approach

Prevention systems, regardless of jurisdiction, are based on a risk-based approach, where the goal is not to eliminate all risks (which is often unfeasible), but to reduce them as low as reasonably practicable (ALARP).

Key conceptual definitions widely used across regulatory and professional literature include:

- Hazard: A source or situation with the inherent potential to cause harm.
- Risk: The probability and severity of adverse effects arising from exposure to a hazard.
- Occupational harm: Any physical or mental outcome resulting from work-related exposure, including accidents, illnesses, fatigue, or psychological strain.

Effective ORP requires moving upstream in the risk chain. That is, not waiting for harm to occur, but acting early, systematically, and strategically. Prevention, in this sense, becomes a continuous improvement process—much like quality or sustainability—and must be integrated into the core operational, managerial, and strategic layers of the organization.

This integration demands coordination among three foundational pillars:

1. Technical-scientific knowledge: Evidence-based risk analysis and control.

2. Legal and regulatory compliance: Enforcement of national and international standards.
3. Organizational and cultural alignment: Shared values and behavior that prioritize safety.

Only through the intersection of these elements can a true culture of prevention emerge—one where safety is not simply enforced from above, but internalized across all levels of the workforce.

2.1.3 ORP as Strategic Value: From Compliance to Competitive Advantage

In today's volatile and high-stakes work environments, ORP has evolved from being a cost or legal burden to becoming a strategic enabler for organizations (Calvente Menéndez, 1996; Del Val y de la Fuente, 1996):

- It reduces absenteeism and turnover, improving workforce stability.
- It lowers insurance premiums and legal exposure.
- It fosters reputation and stakeholder trust.
- It enhances operational continuity, especially in high-risk sectors like construction, mining, transport, or energy.

For digitally transformed organizations, ORP also intersects with the implementation of intelligent systems: from AI-based video analytics for detecting unsafe behaviors, to wearables that monitor fatigue or stress in real time. These technologies, however, must be implemented with ethical rigor, respecting privacy, autonomy, and equity.

2.1.4 A Universal Right, a Shared Responsibility

Ultimately, occupational risk prevention is a universal human right and a shared responsibility. Governments must establish clear and enforceable regulations. Companies must integrate prevention into their strategic and operational decisions. Workers must be empowered and trained to participate meaningfully in their own protection. And researchers and technologists must design solutions that are inclusive, explainable, and accountable.

When properly understood and applied, ORP is not a bureaucratic formality, but a lever of safety, resilience, justice, and human dignity in the world of work. It is the ethical foundation of a sustainable future of labor—one where people are not only productive, but protected, respected, and valued.

2.2 Fundamental Principles of Occupational Risk Prevention

Occupational Risk Prevention (ORP) is not an improvised or discretionary activity. Rather, it is governed by a set of structured, legal, and methodological principles that guide the systematic identification, evaluation, mitigation, and, where possible, elimination of work-related risks. These principles are not mere guidelines; they are the core obligations established by international law—especially the European Framework Directive 89/391/EEC, adopted in 1989—which provides a common foundation for occupational safety and health policies across the European Union, and influences standards worldwide.

The Directive's Article 6 establishes a series of General Principles of Prevention, which have since become a global benchmark. Their influence extends to national legal frameworks, such as ISO 45001:2018, and they serve as foundational criteria for all risk management strategies in the workplace—regardless of sector, geography, or size of the organization.

Below, we examine each of these principles in depth, highlighting their technical scope, logical sequencing, and their practical implications for effective occupational safety management.

1. Avoiding risks: The first and most decisive principle is also the most ambitious: eliminate hazards altogether. If a process, material, or task is inherently dangerous, the most effective measure is to remove it from the workplace entirely.

   This principle reflects a paradigm shift: rather than managing risk, it encourages organizations to design it out. This may involve eliminating hazardous procedures, substituting harmful substances, or redesigning workflows and layouts. It represents the highest level of protection and should always be prioritized in prevention strategies.

2. Evaluating the risks that cannot be avoided: When complete elimination is not possible, a systematic and context-specific risk assessment must be conducted. This includes analyzing both the likelihood of occurrence and the severity of potential harm, considering the specific environment, work tasks, worker profiles, and exposure times.

Risk assessments must be documented, regularly updated, and adapted to dynamic workplace conditions, forming the empirical basis for selecting and justifying appropriate control measures.

3. Combating risks at the source: Prevention should begin where the risk originates. This may involve redesigning equipment, installing safety barriers at machinery, reorganizing workflows to reduce exposure, or isolating hazardous processes. The principle demands technical interventions at the point of risk generation, rather than relying on behavioral corrections downstream.

   This "upstream control" approach reflects the hierarchy of controls promoted by global standards such as NIOSH, OSHA, and ISO 45001, reinforcing that the most effective controls are those that address risk at its origin.

4. Adapting the work to the individual: This principle embodies a human-centered approach to safety. It implies adjusting workstations, tools, and tasks to the physical and cognitive capacities of workers, considering diversity in age, gender, training, and functional ability.

   It also includes mitigating monotonous or repetitive work and promoting ergonomic design to reduce cumulative trauma, fatigue, and psychological strain. This principle aligns with inclusive workplace design, universal accessibility, and equity in work environments.

5. Taking account of technical progress: Prevention must be dynamic and responsive to scientific and technological developments. As new tools, materials, or methodologies become available, organizations are obliged to revise and upgrade their prevention measures accordingly.

   This principle links directly to continuous improvement models and underpins the alignment of ORP with innovation, digitalization, and the use of emerging technologies like AI, IoT, and wearables, particularly in high-risk sectors such as construction or manufacturing.

6. Replacing the dangerous with the less dangerous: Known as the Substitution Principle, this guideline urges the replacement of hazardous agents, processes, or equipment with safer alternatives—when technically and operationally feasible.

   For instance, replacing solvent-based products with water-based ones, or using automated systems to perform high-risk tasks. Importantly, this principle supports sustainable prevention, as it encourages long-term safety enhancements that often lead to better environmental and health outcomes.

7. Developing a coherent overall prevention policy: Risk prevention should not be an isolated activity, but an integrated management function. This principle requires preventive measures to be planned systematically, combining technical solutions with organizational strategies and social considerations.

   This includes integrating ORP into production planning, facility layout, resource allocation, and communication flows. It is particularly relevant in complex systems such as construction sites or manufacturing plants, where multiple actors and variables interact continuously.

8. Giving priority to collective protection over individual protection: Collective protection measures (e.g., guardrails, ventilation systems, automatic shut-offs) must be prioritized over individual ones (e.g., helmets, gloves, masks). This is because collective controls:

   - Protect multiple people at once
   - Are less dependent on human compliance
   - Offer more sustainable risk reduction

   The principle is grounded in effectiveness and resilience, and it aligns with the global Hierarchy of Controls framework used across international safety systems.

9. Giving appropriate instructions to the workers: No prevention system can function without clear, accurate, and continuous communication. Workers must be adequately trained, informed, and engaged in safety processes to ensure they understand the risks and know how to act.

   This principle promotes the participatory dimension of prevention: empowering workers through training, dialogue, and feedback channels. It reflects the broader concept of safety culture, in which behavior, attitudes, and shared values reinforce prevention from within.

The principles above are not isolated checkboxes. They must be applied in logical and hierarchical sequence, starting with elimination, followed by risk assessment, source control, substitution, and so on.

This logic is operationalized through:

- Risk assessment matrices

- Work design criteria
- Preventive planning documents
- Equipment selection protocols
- Procedural workflows and audits

Moreover, these principles must be embedded from the design phase, not simply added as corrective actions after systems are already in place. This is the essence of preventive engineering and design thinking in safety: anticipating risk before it materializes, and shaping systems that are intrinsically safer.

While derived from European regulation, these principles are internationally recognized. Frameworks such as:

- ILO Recommendation No. 197
- ISO 45001:2018 Clause 8.1.2
- ILO-OSH 2001 Guidelines

all echo and expand upon these foundational ideas. Their universal nature allows for local adaptation without compromising on rigor or ambition.

They remain the cornerstone of effective occupational safety and health systems—whether in Europe, Latin America, Asia-Pacific, or Africa—and form the methodological bridge between compliance, performance, and ethical responsibility.

2.3 Areas of Action in Occupational Risk Prevention

Occupational risk prevention extends transversally across all factors that influence the safety and health of individuals within the work environment. This cross-cutting nature translates into a wide range of areas of action, encompassing everything from the material conditions of the physical environment to the organizational and psychosocial factors that shape human behavior.

Identifying these areas is not a theoretical exercise—it is the essential first step toward any effective risk assessment. Only a comprehensive approach makes it possible to build a genuinely operational and responsive prevention system tailored to the complexity of today's workplace.

Effective prevention is not achieved by addressing isolated risks in a fragmented way. Rather, it requires developing a holistic vision that considers all dimensions of the work environment (see Table 6). Only in this way can one design truly adaptive, sustainable interventions capable of protecting workers' health in a real and continuous manner.

Table 6. Preventive Action Areas According to Each Dimension of the Work Environment (table by the author).

Area of Intervention	Examples of Risks	Main Preventive Strategies
Safety Conditions	Cuts, entrapments, falls	Signage, protective devices, equipment maintenance
Environmental Conditions	Noise, pollutants, inadequate lighting	Technical measurements, ventilation, personal protective equipment (PPE)
Physical Workload	Overexertion, awkward postures	Ergonomics, mechanical aids, task rotation
Mental Workload	Stress, sustained vigilance	Task review, scheduled breaks, organizational support
Organizational Factors	Excessive pace, lack of control	Work planning, autonomy, healthy leadership
Psychosocial Risks	Conflict, harassment, uncertainty	Psychosocial assessment, mediation, preventive culture

2.3.1 Safety Conditions

This domain encompasses all material and technical elements which, by their nature or usage, may pose a hazard to workers. It includes:

- Machinery: risks of entrapment, cuts, ejected objects, or malfunction of emergency systems.
- Hand or power tools: design flaws, poor maintenance, or improper handling.
- Installations: electrical networks, ventilation systems, and hazardous substance storage.
- Transit areas: corridors, staircases, and access points, with potential for falls, trips, or collisions.

Preventive strategies in this area focus on safe design, preventive maintenance, compliance with technical specifications and regulatory standards for equipment usage, and proper signage.

2.3.2 Environmental Conditions

This refers to the physical, chemical, and biological factors present in the work environment that can disrupt physiological balance or impact health. Key examples include:

- Noise: continuous or peak exposure exceeding permissible levels.
- Lighting: insufficient, excessive, or unevenly distributed light, which can affect vision and cause fatigue.
- Temperature and humidity: extreme or inadequate thermal conditions for the task performed.
- Chemical contaminants: gases, vapors, dusts, fumes, or toxic/irritant liquids.
- Biological agents: viruses, bacteria, fungi, or parasites encountered in specific sectors (e.g., healthcare, agriculture, cleaning).

Prevention in this domain is grounded in environmental monitoring, proper handling protocols, hygiene and sanitation controls, ventilation systems, and the use of personal protective equipment (PPE).

2.3.3 Workload (Physical and Mental)

Workload represents the combination of physical and cognitive demands imposed on the worker during the execution of their tasks. A disproportionate, poorly managed, or sustained workload can lead to fatigue, musculoskeletal disorders, stress, or functional deterioration.

- Physical workload: manual handling of loads, awkward postures, repetitive movements, or intense exertion.
- Mental workload: information overload, rapid decision-making, sustained vigilance, or multitasking.

Intervention requires ergonomic assessments, task redesign, job rotation, scheduled breaks, and adjustment of work pace to individual and operational capacities.

2.3.4 Organizational and Psychosocial Factors

This area addresses the intangible components of the workplace environment, such as hierarchical structures, corporate culture, management methods, and interpersonal relationships. Factors such as time pressure, role ambiguity, isolation, lack of support, or conflicts between personal and organizational values can give rise to psychosocial risks, potentially leading to chronic stress, burnout, anxiety, depression, or low motivation.

- Time management: shift work, night shifts, overtime, rotation schedules
- Emotional demands: public-facing roles, decision-making with high-stakes consequences.
- Interpersonal relationships: leadership quality, internal communication, conflict resolution mechanisms.

Prevention in this area involves specific tools and strategies: psychosocial risk surveys, structured interviews, organizational culture interventions, training in relational skills, and the redesign of management systems.

2.4 Risk Assessment

Occupational risk assessment is one of the foundational pillars of any workplace health and safety management system. It is a technical, systematic, and documented procedure that enables the identification of hazards in the work environment, the analysis of their nature, the estimation of the associated risk level, and the determination of appropriate preventive or corrective measures. Its objective is not merely to fulfill a legal obligation but to ensure a rational and effective management of the risks that affect the safety and health of workers.

This process is not static: it must be reviewed periodically and, in particular, whenever there are changes in production processes, the introduction of new technologies, reorganization of work, onboarding of new personnel, occurrence of incidents, or updates in applicable regulations.

Beyond its regulatory mandate, risk assessment forms the technical basis upon which the entire preventive system is built. It enables evidence-based decision-making, efficient resource allocation, and provides the technical justification for actions taken. A poorly executed, incomplete, or outdated risk assessment not only endangers workers but also compromises the legal, operational, and reputational integrity of the organization.

In essence, risk assessment is not a mere bureaucratic exercise. It is a critical technical activity that transforms uncertainty into applied knowledge and elevates prevention to a strategic, rational, and adaptive function within the organization.

1. Hazard identification: The first phase consists of systematically and exhaustively identifying all sources with the potential to cause harm. This involves analyzing the physical environment, organizational structure, human interactions, equipment and machinery, substances used, and any other elements capable of causing injury or illness.

   It is crucial not to focus solely on the most visible or well-known hazards: latent, emerging, or underestimated risks must also be considered, including psychosocial hazards and those arising from the use of intelligent technologies.

2. Determination of exposed workers: Once hazards are identified, the next step is to analyze who may be exposed, under what conditions, and with what frequency. This requires evaluating workstations, assigned tasks, shift patterns, and individual characteristics such as age, experience, health status, and gender, as well as identifying potentially vulnerable or high-risk groups.

3. Risk estimation: Probability × Consequence. Risk analysis is typically performed using qualitative, semi-quantitative, or quantitative methods. These involve estimating the likelihood of an incident occurring and the severity of its potential consequences. By combining these factors, a risk level can be determined, allowing the organization to prioritize its preventive actions.

   Common tools include risk matrices, severity scales, weighted indices, and sector-specific methodologies (e.g., INSST, Fine, FMEA, AMFE), adapted to the particular nature of each activity or industry.

4. Proposal of preventive and corrective measures: Once the risk level is determined, it is necessary to define the actions required to eliminate or reduce the risk to an acceptable level, following a hierarchical intervention model:

   - Eliminate the hazard at its source.
   - Substitute hazardous processes or materials.
   - Implement collective technical protection measures.
   - Apply organizational measures (e.g., procedures, shift rotation).

- Use personal protective equipment (PPE) as a last resort.
- Provide training and information to all affected personnel.

These measures must be effective, proportionate to the level of risk, and technically feasible. Their implementation should be fully integrated into the broader preventive planning framework of the organization.

5. Documentation, review, and continuous improvement: The entire risk assessment process must be documented, detailing the methodologies applied, the results obtained, the measures adopted, and the persons responsible for implementation and follow-up. This documentation is not static—it must be reviewed and updated under the following circumstances:
   - When there are material or functional changes in the workplace.
   - When deficiencies or incidents reveal failures in the preventive system.
   - When new technical, regulatory, or medical information becomes available.
   - Periodically, as part of the organization's continuous improvement cycle.

Ultimately, risk assessment is the cornerstone upon which an effective, accountable, and adaptive prevention system is built. It enables informed decision-making, strategic resource allocation, and technical legitimacy of preventive actions. A deficient or outdated risk assessment not only undermines the protection of workers but can also result in regulatory non-compliance and reputational damage.

Thus, risk assessment must be understood as a rigorous and indispensable technical activity—one that transforms ambiguity into operational clarity and embeds preventive culture into the core of organizational strategy.

2.5 Preventive Measures

Preventive measures represent the technical and organizational response to the risks identified during the risk assessment process. Their primary function is to eliminate the hazard or, if elimination is not feasible, to reduce the associated risk to acceptable levels, thereby ensuring safe and healthy working conditions. These are not generic or arbitrary measures; they must be specific, prioritized, proportionate to the level of risk, and tailored to the particular environment and task.

In alignment with European regulations, national legislation—such as that in Spain—stipulates a core principle: preventive measures must be integrated into all business activities. They are not to be implemented as isolated or ad hoc actions.

1. Technical measures: These interventions aim to act directly at the source of the risk, targeting installations, equipment, materials, environments, or production processes. Their purpose is to control the hazard from a physical or engineering standpoint.

   Common examples of technical measures include:

   - Installing mechanical guards on machinery.
   - Substituting hazardous substances with safer alternatives.
   - Signposting risk areas.
   - Using localized extraction systems for contaminants.
   - Ergonomic redesign of workstations.

   Technical controls should always represent the first line of defense, as they offer stable protection that is less dependent on human behavior.

2. Organizational measures: Organizational measures seek to modify the structure, timing, and functional logic of work in order to reduce exposure to risk or prevent its emergence altogether.

   Typical examples include:

   - Task rotation to reduce physical or mental workload.
   - Reorganization of shifts or introduction of active breaks to prevent fatigue.
   - Controlled access to high-risk areas.
   - Implementation of standard operating procedures (SOPs).
   - Integration of risk prevention into operational planning.

   These strategies allow for effective mitigation of factors such as time pressure, cognitive overload, multitasking, and operational disorder.

3. Training and information measures: These measures focus on enabling personnel to understand hazards and adopt safe behavior, fostering a strong preventive culture. They are essential to ensure that all other preventive actions are effective.

   Examples include:

   - Task-specific training on handling hazardous equipment or substances.
   - Clear communication about risks associated with specific job roles.
   - Emergency drills and hands-on safety exercises.
   - Safety manuals and hazard data sheets.

   Such measures must be adapted to the worker's profile (language, education level, experience) and kept up to date.

4. Health surveillance: Occupational health surveillance is a tool for the early detection of adverse effects resulting from work-related exposures and for verifying the effectiveness of the preventive system. It must be conducted in accordance with principles of medical confidentiality, worker consent (except in legally justified exceptions), and relevance to specific workplace risks.

   It includes:

   - Initial, periodic, and post-absence medical examinations.
   - Biological monitoring of exposure to chemical or physical agents.
   - Programs for tracking occupational diseases.
   - Individual job adaptations based on health surveillance findings.

   Importantly, health surveillance is not a substitute for risk control. Its role is complementary, contributing to internal epidemiological monitoring and prevention enhancement.

The preventive principles outlined in section 2.2 apply here as well. Preventive measures should follow a hierarchical logic based on structural effectiveness:

1. Eliminate the hazard.
2. Substitute hazardous agents or processes.
3. Control risks at the source with technical measures.

4. Apply collective protection measures.
5. Use personal protective equipment (PPE) as a last resort.
6. Provide training and information to workers.
7. Monitor performance and correct deviations.

The earlier the intervention in the risk chain (i.e., at the source), the more effective, cost-efficient, and sustainable the preventive strategy will be.

Preventive measures must not be implemented as standalone solutions. They must form part of a coherent and integrated preventive management system. This entails:

- Their inclusion from the design phase of work processes.
- Alignment with the risk assessment outcomes.
- Allocation of sufficient resources.
- Establishment of verifiable and measurable objectives.
- Understanding and acceptance by the workforce.

Only under these conditions can prevention move from being reactive and fragmented to becoming a strategic, anticipatory, and structural function within the organization.

2.6 Preventive Culture

Preventive culture represents the most mature and transversal level of integration of occupational risk prevention within an organization. It goes far beyond legal compliance or the formal implementation of technical measures: it constitutes a way of thinking, acting, and managing work where the health and safety of people are seen as fundamental and collectively shared values.

This cultural approach transforms prevention into a structural component of organizational identity, influencing both strategic decisions and everyday behaviors. In organizations with a robust preventive culture, safe behaviors are not merely the result of rules or external oversight—they emerge organically and are sustained across all hierarchical levels.

According to the European Agency for Safety and Health at Work (EU-OSHA, 2018), preventive culture refers to the set of beliefs, attitudes, perceptions, values, and behaviors shared by members of an organization regarding the management of occupational risks.

In other words, it is the organizational climate that determines how safety is prioritized, perceived, and managed—not just what is written in safety manuals, but what is actually practiced, tolerated, or enforced in the daily work environment.

Organizations with a strong preventive culture exhibit distinctive and recognizable traits (EU-OSHA, 2018; Schor-Landman, 2022):

- Visible leadership commitment: Senior management does not merely endorse prevention plans—they embody them through example, allocate dedicated resources, and embed safety into the strategic core of the organization.
- Active worker participation: Workers are encouraged to engage in horizontal dialogue, submit suggestions, and contribute to risk identification and continuous improvement.
- Open and two-way communication: Information on risks, procedures, and corrective actions is shared clearly, consistently, and transparently.
- Positive response to errors and incidents: Events are investigated without seeking blame, with a focus on organizational learning and system improvement.
- Integration of safety into decision-making: Safety considerations are taken into account from the early stages of process, facility, or shift design.
- Continuous, non-formalistic training: The focus is on developing real preventive competencies, rather than merely fulfilling administrative training requirements.

However, building a strong preventive culture is not without its challenges. Common barriers include:

- Reactive and compliance-driven focus: When prevention is reduced to ticking boxes or responding after incidents occur.
- Lack of preventive leadership: A disconnect or low involvement from middle and upper management.
- Prevention–production dichotomy: Viewing safety as a constraint rather than a driver of operational excellence.

- Resistance to change: Entrenched habits that normalize risky practices as part of "how things are done."
- Absence of behavioral evaluation: Failure to monitor or positively reinforce safe behaviors within teams.

Overcoming these barriers requires structural, progressive interventions aligned with the organization's overall strategic vision—not isolated initiatives or temporary campaigns.

A strong preventive culture does more than reduce the frequency and severity of incidents. It also:

- Improves workplace climate and team motivation.
- Enhances trust between leadership and frontline personnel.
- Lowers the hidden costs of unsafe environments (e.g., downtime, rework, staff turnover).
- Boosts the organization's external reputation and stakeholder relations.
- Facilitates the adoption of technological innovations (e.g., artificial intelligence) from a responsible and safe perspective.

In essence, a mature preventive culture is a cornerstone of business sustainability—aligned with the principles of corporate responsibility, operational efficiency, and organizational resilience.

Developing a preventive culture is not about issuing protocols—it is a continuous process of organizational transformation. This process involves:

- Conducting an initial assessment of cultural maturity.
- Providing cross-functional training and modeling leadership behavior.
- Establishing cultural indicators (e.g., risk perception, engagement, trust).
- Regularly evaluating progress and impact.
- Recognizing and celebrating internal best practices.

There are no one-size-fits-all formulas, but there are universal principles: coherence, consistency, participation, and leading by example.

2.7 Current Challenges in Occupational Risk Prevention

Occupational risk prevention (ORP) has undergone significant development since its formal legal consolidation in the 1990s (Pantoja-Rodríguez et al., 2017; Schor-Landman, 2022). However, the productive, technological, and social environment in which it operates has changed drastically. These changes have introduced new structural challenges that condition the effectiveness, applicability, and long-term sustainability of prevention systems (Fernández-Costales Muñiz, 2020).

These challenges are not circumstantial or marginal—they reflect deep transformations in the world of work: new organizational models, accelerated digitalization, diversified risk profiles, and increasingly complex interactions between people, technology, and environment. The ability of occupational risk prevention to adapt to these evolving conditions will define its future relevance as a strategic tool for both management and protection (Fernández-Costales Muñiz, 2020; Ortega Giménez et al., 2021).

1. Growth of temporary work and subcontracting: A major challenge today lies in the fragmentation of employment and the outsourcing of functions. The rise of temporary contracts, cascading subcontracting chains, and the proliferation of economically dependent self-employment hinder:
    - Uniform application of preventive measures.
    - Traceability of responsibilities in the event of an incident.
    - Coherent planning of preventive systems in shared work environments.
    - Equitable access to training, information, and health surveillance.

    This issue is especially critical in sectors like construction, logistics, and auxiliary services, where multiple companies and workforce profiles coexist in the same space.

2. Aging workforce: The extension of working life, coupled with demographic decline, has led to a growing presence of older workers in physically demanding or technologically intensive environments. This demographic trend presents specific challenges:
    - The need to redesign tasks and workstations in line with functional capacity.
    - Increased vulnerability to certain chronic or occupational diseases.
    - Adjustments to work rhythms and recovery times.

- Personalized approaches to health surveillance.

Prevention must evolve toward a more inclusive, ergonomic, and adaptive model that accounts for diverse capacities across the working lifespan.

3. Emergence of new psychosocial risks: Psychosocial risks have taken on increasing importance in recent years. Chronic stress, mental fatigue, workplace harassment, isolation, performance pressure, and organizational ambiguity directly affect mental and physical health. These risks cannot be detected with conventional tools or mitigated through personal protective equipment. They require:
    - Specialized evaluations using validated instruments.
    - Multidisciplinary interventions (psychology, organizational design, leadership).
    - Cultural transformations in managerial styles.
    - Protection from invisible yet cumulative harm.

    Workplace mental health is no longer an ancillary concern—it is a critical indicator of organizational sustainability.

4. Emerging technologies and advanced automation: The arrival of technologies such as artificial intelligence, cyber-physical systems, advanced automation, and smart sensor networks has fundamentally transformed working conditions. These innovations introduce new types of risk not contemplated in traditional risk assessment frameworks, such as:
    - Human–machine interaction in unstructured environments.
    - Cognitive fatigue from information overload.
    - Misalignment between human capabilities and technological demands.
    - Functional safety issues arising from algorithmic failures or autonomous decisions.

    Occupational risk prevention must expand its conceptual and technical frameworks to address these emerging technological risks—without losing sight of the human-centered nature of prevention systems.

5. Supervision of complex and dynamic environments. In sectors such as construction, port logistics, or industrial maintenance, workplaces are highly dynamic, multi-actor, and continuously evolving. This complicates:

   - The effective implementation of planned safety measures.
   - Real-time monitoring of changing conditions.
   - Coordination between concurrently operating contractors or service providers.
   - Availability of accurate, up-to-date risk information on-site.

   These settings demand more agile, sensor-based, interoperable prevention models capable of rapid response.

6. Inequality in access to prevention: There is a real gap between large companies (which often operate with in-house prevention services) and small businesses or self-employed workers (who rely on third-party services and often perceive prevention as an administrative burden). This creates a structural inequality in worker protection that may result in:
   - Significant differences in hazard exposure.
   - Limited access to preventive innovation.
   - Weak safety culture at operational levels.

   Bridging this gap requires proactive public policy, targeted incentives, and accessible tools—particularly for micro-enterprises and vulnerable sectors

The challenges currently facing occupational risk prevention cannot be resolved solely by updating regulations or improving procedures (Fernández-Costales Muñiz, 2020; Schor-Landman, 2022). They call for:

- Rethinking prevention from a systemic perspective that accounts for evolving work models and social transformations.
- Integrating technology as an enabler—not an end in itself—while upholding the dignity and centrality of human workers.
- Bringing new professional profiles into the field (e.g., occupational psychologists, data analysts, inclusive designers).
- Shifting from reactive prevention to proactive, anticipatory safety monitoring based on data, sensors, and predictive analytics.

This strategic reorientation is essential for ensuring that occupational risk prevention remains a relevant, resilient, and transformative component of future work environments.

3 General Applications of AI in Occupational Risk Prevention

3.1 Introduction

Artificial intelligence is redefining traditional paradigms of occupational risk prevention. Its ability to analyze large volumes of data, identify complex patterns, process real-time information, and anticipate critical events positions this technology as a first-tier strategic tool for enhancing workplace health and safety (Rodríguez Elizalde, 2025).

Unlike classical approaches—which tend to be reactive, reliant on manual supervision, and limited by human analytical capacity—AI systems enable the automation of monitoring processes, the anticipation of risky behaviors, and the optimization of preventive decision-making with increasing precision (Goñi Sein, 2024; Macías García, 2023). AI does not replace core preventive principles; rather, it strengthens and expands them, integrating seamlessly into various levels of occupational safety management systems.

In an environment marked by increasingly complex, dynamic, and technologically intensive workplaces, artificial intelligence offers solutions to (Macías García, 2022; Nájar Becerra, 2024) (Fig. 3):

- Detect hazardous conditions in real time using smart sensors and computer vision.
- Analyze historical accident patterns through machine learning algorithms.
- Automate training, inspection, and documentation processes, reducing the operational burden on the safety system.
- Customize control measures based on individual risk profiles and the evolving characteristics of the work environment.

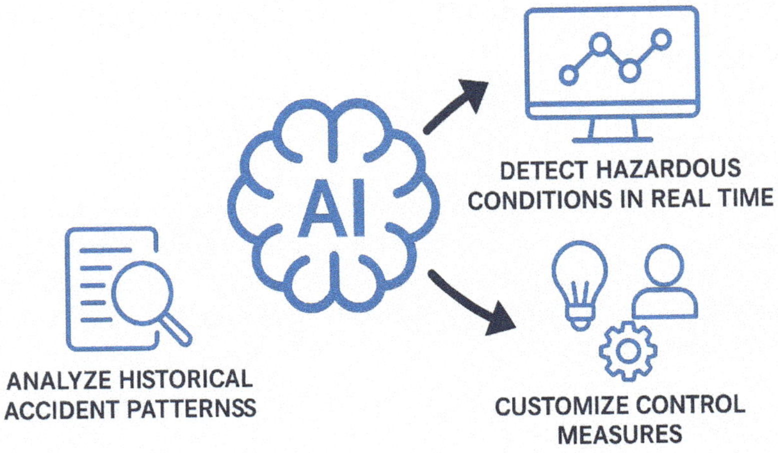

Fig. 3. A Cross-Cutting Technology for a Changing Environment (graphic by the author).

This approach transforms occupational risk prevention from a reactive, compliance-based function into an anticipatory, adaptive strategy grounded in objective evidence (Ruiz & Velásquez, 2023).

Artificial intelligence solutions do not merely observe: they learn, correlate, predict, and act. This capability enables the deployment of new models of prevention (Goñi Sein, 2024):

- Predictive prevention: anticipating the emergence of hazardous conditions before they materialize (Macías García, 2022).
- Context-aware prevention: adapting preventive responses to contextual variables such as location, environmental conditions, or worker profile (Nájar Becerra, 2024).
- Continuous prevention: operating in real time and without interruption (Macías García, 2023; Rodríguez Elizalde, 2025).
- Scalable prevention: applicable across multiple worksites, shifts, or concurrent processes (Otero Mateo et al., 2023).

Moreover, the integration of artificial intelligence with technologies such as the Internet of Things (IoT), Big Data, and Building Information Modeling (BIM) significantly amplifies its potential—creating intelligent work environments capable of self-adjusting to dynamic risk conditions (Morales-Conejo, 2021; Rico-Bautista et al., 2021; Tito Chura, 2023).

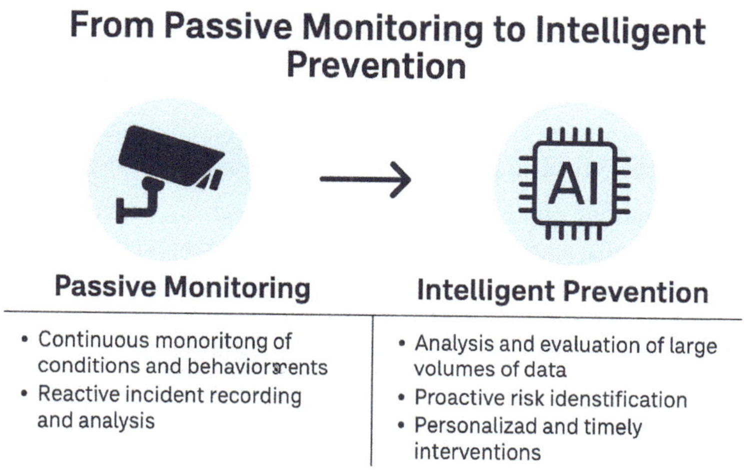

Fig. 4. From Passive Monitoring to Intelligent Prevention (graphic by the author).

The incorporation of artificial intelligence into the field of occupational risk prevention is not merely a technical matter; it demands a rethinking of supervision models, the assignment of responsibilities, data confidentiality, and professional training (Chicaiza Ortiz, 2024; Goñi Sein, 2024; Rodríguez Sáenz de Galdeano, 2024). Algorithms must be auditable, explainable, and socially acceptable—especially when they influence decisions that directly impact people's safety (Aránguiz Villagrán, 2022; Lassi, 2023).

Therefore, the adoption of these technologies must be accompanied by:

- Rigorous technical assessment of their applicability and reliability (Rojas-Gualdron, 2022).
- Ethical design centered on the protection of individuals (Rodríguez-Martínez et al., 2023).
- Proper training for occupational health and safety professionals in their use and interpretation (Estardid Colom, 2023).

The following sections present the most relevant applications of artificial intelligence in occupational risk prevention, organized by strategic function:

monitoring, predictive analysis, computer vision, virtual assistance, psycho-physiological evaluation, automated auditing, and integration with digital systems (Chicaiza Ortiz, 2024). Each application is described from both a technical and functional perspective, illustrating its contribution to designing safer, more resilient, and smarter workplaces.

3.2 Real-Time Monitoring

Real-time monitoring powered by artificial intelligence is transforming how organizations approach occupational risk prevention—particularly in sectors like construction, where work environments change continuously and risks can materialize in a matter of seconds (Macías García, 2023). This technology enables the detection of hazardous conditions at the very moment they arise (Aránguiz Villagrán, 2022; Macías García, 2022), triggering immediate alerts or even activating automated responses without the need for direct human intervention.

These systems no longer merely record data; they interpret, learn, and act—turning the workplace into an environment capable of monitoring its own safety status (Goñi Sein, 2024).

An intelligent monitoring system operates in real time through an ecosystem of interconnected technologies. Here is how it is typically structured:

1. Internet of Things (IoT) Sensors (Campero-Jurado et al., 2020): These continuously collect data on gases, temperature, humidity, vibrations, noise, or movement (Luna López et al., 2021). They are embedded in helmets, tools, or distributed across the worksite.
2. Computer Vision: Smart cameras recognize people, objects, and risky situations (Márquez-Sánchez et al., 2021), such as missing helmets, entry into restricted areas, or imminent falls.
3. Communication Networks: These transmit data instantly to analysis platforms using Wi-Fi, 5G, or industrial communication networks.
4. AI-Based Analytics Platform: This system processes incoming data, identifies anomalous patterns, predicts dangerous situations, and executes decisions based on predefined safety protocols (Campero-Jurado et al., 2020; Márquez-Sánchez et al., 2021).
5. Response System: It issues alerts via mobile devices, wearables, or audible signals. In some cases, it can even shut down machinery or restrict access

automatically (Galván Guerra et al., 2024; Ramírez Véliz et al., 2022; Rojas-Gualdron, 2022).

Through this structure, the work environment becomes an active sensory system—capable of assessing its own safety status and responding to risk before harm occurs.

One of the most effective and visible applications of computer vision in construction sites is the automated control of personal protective equipment (PPE) compliance (Massiris et al., 2021; Rodríguez Elizalde, 2025). Using cameras and image recognition algorithms, the system can automatically detect whether a worker is wearing a helmet, reflective vest, or harness. If not, an immediate alert is triggered. Cameras positioned at entry points, corridors, and critical areas capture real-time footage, which is analyzed by deep learning algorithms specifically trained to recognize safety helmets, high-visibility vests, harnesses, or gloves (Delhi et al., 2020).

Fig. 5. Example of PPE Detection Using Computer Vision (courtesy of Seitech).

When the system detects that a worker enters a site without the required personal protective equipment (PPE), or that the equipment is incorrectly worn (e.g., an unfastened helmet), it automatically generates an alert that can be sent directly to the worker, a safety officer, or even used to deny access to the area (Massiris et al., 2021; Ruiz & Velásquez, 2023).

This solution reduces reliance on human visual inspections, prevents the normalization of unsafe behavior, and acts proactively—before actual exposure to risk occurs (Moreno Rabel & Fernández Muñoz, 2022; Rodríguez Elizalde, 2025).

One of the most prominent examples in this field is the Calidus system. Developed by Seitech (Seitech Solutions, n.d.-a), Calidus is a platform specialized in automating occupational risk prevention using computer vision. Its main objective is to ensure proper use of PPE by workers—particularly in remote or hard-to-monitor environments such as linear construction sites.

It is, therefore, a smart and modular platform that integrates computer vision, remote supervision, and on-site control to systematically, traceably, and in real time, verify that workers comply with the company's safety protocols.

Calidus functions as a configurable suite of services and technological tools that allows companies to:

1. Automate PPE compliance checks.
2. Monitor field operations remotely.
3. Verify preventive measures through on-site validation.

Thanks to its modular architecture, companies can tailor the solution to match their internal occupational health and safety protocols, enhancing preventive control without increasing the operational workload for supervisors or safety managers.

One of Calidus's most powerful components is its computer vision system, developed in collaboration with Vision Analytics. This system automates the detection of PPE compliance using images captured via mobile devices, desktops, or tablets. Through AI algorithms trained according to each company's specific safety criteria, the system can:

- Detect whether workers are correctly wearing mandatory PPE for each task and activity (e.g., helmets, reflective vests, harnesses, gloves, safety glasses).

- Verify that the technician is carrying required equipment in their vehicle (first-aid kit, ladders, toolboxes).
- Generate alerts when minimum safety conditions are not met.
- Log all information in real time on a traceable and auditable platform.

This feature is particularly valuable in decentralized operations, where continuous on-site supervision is not feasible. The analysis is immediate, requires no human intervention, and the records are securely stored on the platform for audit, inspection, or legal compliance purposes.

In many field operations, unexpected issues, uncertainties, or real-time validation needs may arise. To address these, Calidus incorporates a real-time communication tool with video and bidirectional audio support, which enables

- Remote assistance from a supervisor during task execution.
- Live verification of safety conditions.
- Instant resolution of doubts or discrepancies.
- Capture of audiovisual evidence for recordkeeping or training purposes.

This direct channel not only boosts operational efficiency but also ensures a quicker and more effective preventive response. It allows field technicians to feel supported in real time—without having to wait for a physical inspection.

The high-quality video and audio features enable precise remote visual inspections, making it possible to perform technical evaluations in remote or restricted-access locations. Without a doubt, this capability represents a significant technical advantage.

In addition to its automated and remote capabilities, Calidus incorporates an on-site supervision service carried out by qualified Seitech technicians, who act as external auditors of compliance with occupational health and safety protocols. These visits can be scheduled periodically or requested on demand, and they serve to:

- Validate that tasks are performed in accordance with the established safety protocols.
- Detect deviations that may not be visible to automated systems.
- Conduct interviews with personnel to assess their understanding of preventive measures.

- Issue reports with specific improvement recommendations.

This in-person component introduces a qualitative control dimension, based on technical expertise and professional judgment, which complements the automated mechanisms. By integrating human oversight into the system's architecture, the evaluation becomes more contextualized and sensitive to nuances that algorithms are not yet capable of accurately interpreting. Moreover, the active presence of technical staff on site reinforces the organization's commitment to safety, promoting a visible and credible preventive culture in line with the principles of proactive occupational risk management leadership.

In 2022, the company Circet España conducted a pilot project with the Calidus platform developed by Seitech at its Seville branch (Interempresas, 2022a; Seitech Solutions (eFSO), 2023a). The aim of this pilot was to validate the potential of artificial intelligence and computer vision in automating compliance monitoring of occupational risk prevention protocols—especially regarding proper use of personal protective equipment (PPE).

The system, configured according to the company's technical requirements, was able to automatically and in real time detect whether workers were correctly wearing items such as chinstrap-fastened helmets, high-visibility vests, safety goggles, gloves, and harnesses. Additionally, Calidus was used to carry out smart inventory checks of safety equipment in technicians' vehicles and to verify the proper use of ladders and other auxiliary gear.

The results of the pilot were highly satisfactory in terms of accuracy, operational efficiency, and compliance traceability. The platform proved to be an effective tool to strengthen preventive culture and optimize resources allocated to field safety supervision—especially in decentralized operations. As a result, Circet decided to scale the implementation nationwide, benefiting over 1,500 technicians. The integration of Calidus has enabled the organization to maintain a systematic and traceable record of compliance with its safety policies, consolidating a prevention model based on evidence, automation, and continuous improvement.

This case clearly illustrates the tangible impact that artificial intelligence, when tailored to the operational needs of the technical-logistics sector, can have on preventive risk management.

Hazardous environmental conditions are a defining factor in many construction sites. Moreover, these conditions can change abruptly, altering the level of risk to which workers are exposed. Indeed, variables such as wind, temperature, dust concentration, or gas accumulation can create dangerous situations—especially in confined spaces, tunnels, or excavation zones.

By installing appropriately placed sensors throughout the site, it becomes possible to monitor levels of toxic gases, extreme heat, and, for example, ambient noise. When these readings exceed predefined threshold values, response protocols are automatically triggered, prompting immediate actions such as increased ventilation or evacuation procedures.

This type of system is already in use through real-time connected atmosphere sensors in underground works or in operations inside tanks, silos, or technical chambers. The use of Internet of Things (IoT) sensors has proven highly effective in monitoring critical variables such as:

- Presence of combustible or toxic gases (e.g., CO, H_2S, CH_4)
- Ambient temperature and relative humidity
- Noise and vibration levels
- Concentration of respirable particulates (PM2.5, PM10)

This real-time environmental and atmospheric monitoring allows workers to be protected against invisible and highly dangerous hazards, such as intoxication or heatstroke, which in most cases cannot be perceived in time by human senses.

Another area where real-time monitoring has proven highly effective is personnel access control and the enforcement of restricted zones within a construction site. Certain areas must be access-restricted due to high-risk conditions (Fig. 6), such as:

- Elevated structures lacking collective fall protection,
- Heavy equipment operation zones or machinery swing radii,
- Confined spaces,
- Areas under suspended loads,
- Proximity to overhead power lines, among others.

Fig. 6. Simulation of a warning triggered by unauthorized access to a restricted transit zone (graphic by the author).

An intelligent access control system based on geofencing and proximity sensors can automatically detect when a worker enters an unauthorized area. Using technologies such as RFID, BLE beacons, or GPS embedded in the worker's helmet, vest, or wristband, the system cross-references the worker's real-time location with the digital site layout (potentially integrated into BIM). If access occurs without the required personal protective equipment or without prior authorization, an alert is triggered—or, in some cases, a physical barrier (such as automated doors or turnstiles) is activated to block passage.

This enhances control over dynamic environments, reduces unintentional human error, and enables tracking and analysis of personnel movements to improve preventive planning—ultimately lowering the incidence of accidents caused by disorientation or procedural oversights.

Monitoring the physical condition of workers is another critical factor. Physical and mental fatigue is among the most common causes of workplace accidents in construction, particularly during long shifts, in hot environments, or while performing monotonous and repetitive tasks. Using smart wearables—such as biometric sensor-equipped wristbands, belts, or vests—key physiological variables can be continuously monitored, including:

- Heart rate
- Core body temperature
- Skin conductivity (an indicator of stress)
- Posture and movement (detectors of instability)

When the system detects a potentially dangerous physiological condition—such as sustained elevated heart rate or signs of fatigue—it can recommend a break, disable a hazardous tool, or alert a supervisor. This helps prevent injuries from overexertion, syncope, heat stress, and exhaustion-related falls, especially during high-risk or critical operations.

Research from Chiba University (Japan) has demonstrated the potential of continuous physiological monitoring as an advanced tool for evaluating attentional state and risk perception in workplace environments (Ashida et al., 2024; Kawasaki et al., 2024). One of the most notable contributions is the use of near-infrared (NIR) imaging techniques to estimate pulse waves non-invasively through facial analysis (Kawasaki et al., 2025). This computer vision–based method is resilient to environmental noise and enables remote, real-time biometric data acquisition without the need for physical contact.

From an occupational safety perspective, these technologies offer an effective way to correlate key physiological parameters—such as heart rate, pulse variability, or facial vascular patterns—with indicators of fatigue, thermal stress, or cognitive decline (Song et al., 2017). Their integration into operational settings allows continuous psychophysiological monitoring of the worker and the early detection of functional deterioration that often precedes critical incidents or errors.

Because they do not require direct contact devices, these solutions are especially useful in mobile, dynamic, or hard-to-supervise environments, such as construction sites, maintenance industries, or confined spaces. Overall, these studies lay the foundation for physiologically aware prevention models, where artificial intelligence and remote sensing technologies are combined to offer more precise, context-sensitive, and person-centered control.

As a complementary element, unmanned aerial vehicles (UAVs)—or drones—equipped with high-definition cameras can be deployed to inspect large elevated areas or zones with limited accessibility or heightened risk (Rodríguez Elizalde, 2022), such as scaffolding, rooftops, or under-construction structures. This avoids exposing additional workers to hazards and allows early detection of anomalies or unsafe working conditions in hard-to-reach places.

The above confirms that intelligent monitoring does more than enhance safety—it transforms the logic of the entire preventive system. Among its most important benefits:

- Faster response times to critical events.
- Enhanced predictive analysis through real-time structured data.
- Proactive prevention: anticipating risks instead of merely controlling them.
- Automated compliance, with instant records and full traceability.
- Integration with other platforms such as BIM, ERP, or maintenance systems.

To ensure that AI-powered monitoring is both effective and ethically sound, several key aspects must be considered:

- Context-aware design: each worksite has unique characteristics.
- Data protection: biometric and visual surveillance must strictly comply with privacy regulations.
- Transparency and explainability: system decisions must be auditable and interpretable.

- Personnel training: operators must understand how the system works, how it responds, and how to act in the event of an alert.
- Interoperability: AI should be integrated into the broader management system—not operate in isolation.

Real-time monitoring powered by artificial intelligence heralds a new paradigm in prevention: situational, personalized, anticipatory, and adaptive. Safety is no longer solely reliant on human supervision but on systems that perceive the environment, interpret it, and respond based on moment-to-moment risk levels

Rather than reacting to accidents after they occur, this approach identifies early warning signs and intervenes before harm is done. This not only protects lives—it enhances operational efficiency, reduces costs, and elevates the preventive maturity of the entire organization.

3.3 Predictive Analysis of Accidents

Predictive analysis stands out as one of the most powerful and transformative applications of artificial intelligence in the field of occupational risk prevention. In fact, it might even be described as intelligent risk anticipation (Macías García, 2022).

Through machine learning techniques, this methodology enables highly accurate forecasting of hazardous events by identifying hidden patterns in large volumes of historical and contextual data—patterns that are impossible to detect through traditional methods (Zamora-Cevallos et al., 2024).

This capability represents a paradigm shift in the logic of prevention: it goes beyond the classical, reactive model based on accumulated experience and ushers in a proactive, dynamic, and evidence-based approach, adjusted to the real context of work activity (Matías et al., 2008).

3.3.1 Stages of Predictive Analysis

The predictive analysis process involves several stages that integrate heterogeneous data, advanced algorithms, and mathematical models capable of detecting complex relationships between variables. The main stages include (Macías García, 2022; Matías et al., 2008; Zamora-Cevallos et al., 2024):

- Collection of multivariable data: Diverse data sources are integrated, such as accident reports, near-miss logs, weather conditions, shift patterns, use of personal protective equipment (PPE), ergonomic variables, occupational health records, and productivity indicators.

- Data preprocessing: Outliers are removed, data are normalized, and qualitative variables are transformed into algorithm-readable formats. This cleaning step is crucial for model reliability.
- Model training: Supervised algorithms—such as logistic regression, decision trees, neural networks, or ensemble models—are applied to labeled datasets (e.g., days with and without incidents) to learn complex causal relationships.
- Validation and tuning: Models are tested with new data to assess predictive accuracy, refining their structure and parameters to reach optimal performance.
- Deployment and operational integration: Once validated, models are implemented for real-time or periodic use, generating preventive alerts, risk heat maps, tailored recommendations, or classifications of critical tasks.

3.3.2 Value of Predictive AI in Risk Prevention

The true value of artificial intelligence lies not in detecting the obvious, but in revealing combinations of seemingly innocuous variables that—when occurring together—significantly raise the probability of an accident (Mosquera et al., 2021). Some illustrative examples:

- Increased fall risk when scaffold use coincides with high wind, evening shifts, and newly hired workers.
- Higher incidence of musculoskeletal injuries during shifts with low task rotation, high humidity, minimal breaks, and cumulative physical load.
- Tendency for errors or near-misses during the final hour on Fridays among teams exceeding 90% of their weekly workload.

These findings do not replace professional judgment—they enhance it with objective, verifiable, and context-aware data.

3.3.3 Practical Applications of Predictive Analysis

Predictive analysis is already being implemented by organizations with advanced safety maturity levels to revolutionize their safety strategy. Notable applications include (Goñi Sein, 2024; Mosquera et al., 2021; Zamora-Cevallos et al., 2024):

- Dynamic risk mapping by activity, location, shift, or worker profile, updated automatically based on climate conditions, machinery presence, or workload. For example, on a large construction site, the system can highlight increased risk in elevated access areas during night shifts with poor weather, prompting targeted action.
- Intelligent task reordering to reduce cumulative exposure to risk factors. If overexertion injuries rise after six hours of repetitive work, the system can adjust shift sequences to prioritize physical tasks earlier in the day.
- Proactive classification of risk profiles to anticipate incidents such as slips, entrapments, or operational errors. For instance, a newly hired worker with no experience in working at height, assigned to an aerial platform during a night shift, could be flagged as a critical profile (see Fig. 7), triggering enhanced supervision and additional training.
- Targeted training for tasks or groups with higher accident probabilities. If a rise in load-handling incidents is detected in certain shifts, the system can automatically launch a tailored training intervention, including AR-based simulations.
- Real-time evaluation of corrective measures, allowing their effectiveness to be verified through predictive safety indicators rather than only after incidents.
- Intelligent prioritization of inspections and audits, focusing resources on areas, teams, or processes with the highest forecasted risk. Rather than planning uniform rounds, the system may recommend immediate auditing in zones where temporary staffing levels are high and outdated machinery is in use.

3.3.4 Strategic Shift Enabled by Predictive Analysis

These use cases demonstrate that predictive analysis is not a post-incident diagnostic tool—it is a strategic anticipation instrument that enables smarter resource allocation, more adaptive planning, and evidence-based preventive decision-making.

Its integration into safety management redefines the role of the safety professional, empowering them with advanced analytics capabilities and delivering a clearer, more precise vision of risk—one that evolves in real time and helps create safer, smarter work environments.

Fig. 7. Inexperienced Worker Assigned to Elevated Platform on Night Shift, Classified as Critical Risk Profile and Assigned Reinforced Preventive Measures (graphic by the author).

A paradigmatic example would be a construction company that, through its predictive system, identifies an increase in falls from elevated platforms on Friday afternoons, particularly under conditions of high humidity and low visibility. Armed with this information, the company can reschedule tasks, limit the use of certain equipment, or reinforce safety supervision measures.

Compared to conventional statistical models—often constrained by fixed assumptions and linear relationships—artificial intelligence algorithms offer several key advantages:

- They learn directly from data, without the need for pre-defined assumptions.
- They detect non-linear relationships, multiple interactions, and temporal patterns.
- They update continuously as new data is incorporated.
- They integrate seamlessly with sensors, monitoring platforms, and safety management systems.

These capabilities make predictive analytics a strategic tool for reducing uncertainty, tailoring preventive actions, and rigorously prioritizing interventions based on technical criteria.

3.3.5 Key Considerations for Successful Implementation

The successful deployment of these technologies depends on several essential factors that ensure effective implementation:

- Data quality and diversity: The model's predictive power depends directly on the richness, representativeness, and reliability of the input data.
- Transparency and ethics: Models must be auditable, understandable, and explainable—especially when they influence decision-making about people or processes.
- Operational integration: Analytical outputs must be translatable into clear, actionable, and practical steps that safety professionals and supervisors can implement.
- Analytical culture: Professionals must be adequately trained to interpret models, validate their outputs, and apply recommendations responsibly and effectively.

3.3.6 From Reactive to Anticipatory Prevention

Predictive analytics powered by artificial intelligence enables a shift from reactive to anticipatory prevention. It doesn't merely record what has happened—it generates actionable insights about what is likely to happen.

It enables organizations to redesign work conditions before harm occurs, using an approach grounded in data, real-world context, and empirical evidence.

This qualitative leap transforms prevention into a function of operational intelligence: an environment in which data not only explain the past but also guide what needs to be done to prevent future incidents. A workplace where safety is not just a goal, but a dynamic, predictive capability embedded at the very core of day-to-day operations.

3.4 Computer Vision for Risk Detection

Computer vision, a branch of artificial intelligence focused on the automated interpretation of images, has become a highly valuable tool in the field of occupational risk prevention. Through smart cameras and deep learning models, this technology enables real-time visual monitoring of the work environment, identifying unsafe conditions, hazardous behaviors, and deviations from established safety protocols (Uribe, 2024a).

This approach represents a significant advancement over traditional passive surveillance. It is no longer just about recording what happens at a worksite or industrial facility, but about understanding and interpreting visual events to generate automated alerts, compliance reports, or immediate responses to critical situations. In essence, cameras become cognitive sensors capable of recognizing patterns, detecting anomalies, and acting autonomously based on the operational context.

A typical computer vision risk detection system usually integrates the following elements (Goñi Sein, 2024; Leguizamón Páez & Becerra Correa, 2024; Moreno Rabel & Fernández Muñoz, 2022):

- Image Capture: Using fixed or mobile cameras (installed on drones, helmets, or vehicles), high-resolution video sequences are captured from critical zones.
- Local or Cloud Processing: The frames are analyzed in real time using algorithms specialized in object recognition, semantic segmentation, and anomaly detection.
- Event Classification: The system differentiates between normal and potentially hazardous situations, such as lack of personal protective equipment, risky postures, deficient signage, or improper material placement.
- Alert Generation: Upon detecting a risk, the system activates an automated protocol which may include notifying a supervisor, creating an event in the safety management platform, or even triggering physical alerts (lights or sounds) in the immediate area.
- Logging and Traceability: Each detected incident is documented with timestamp, geolocation, and visual evidence, facilitating audits and feeding predictive analytics models.

This technology's versatility allows its implementation across a wide range of scenarios, generating concrete benefits in risk anticipation and safety control optimization. Some illustrative use cases include:

- Automated verification of PPE usage: As discussed in section 3.2 (Real-Time Monitoring), vision algorithms can identify in real time whether workers are properly wearing required protective equipment (helmet, vest, gloves, goggles, harness).
- Detection of poorly signaled or disorganized environments: Using image segmentation and object detection, the system identifies obstructions in walkways, misplaced materials, or absent/damaged safety signage. For example, a pallet blocking an evacuation route can be detected on a construction site (Fig. 8), prompting an immediate alert.

Fig. 8. llustration of a pallet placed in a designated evacuation route on a construction site (graphic by the author).

- Recognition of unsafe behaviors: AI models analyze movement patterns and body postures indicating risk exposure, such as bending without knee flexion, improper load handling, or accessing heights without fall protection. These can be identified before an accident occurs, enabling proactive intervention. For instance, if a worker walks under a suspended load during hoisting operations (Fig. 9), the system can trigger an alarm and log the incident.

Fig. 9. Illustration of a worker walking under a suspended load and the simulated warning generated by the application (graphic by the author).

- Occupancy control and safety distance monitoring: With people-counting and tracking techniques, computer vision detects overcrowding or non-compliance with minimum distance requirements in sensitive areas (confined spaces, ATEX zones, electrical risk areas). For example, in an electrical substation, the system may prevent more than two people from entering simultaneously by triggering a visual alert and temporarily locking entry.
- Aerial supervision via drones: As covered in section 3.2, in complex or hard-to-reach environments (scaffolding, roofs, metal structures), drones equipped with cameras and vision algorithms allow for systematic visual inspections without exposing workers. For instance, improper anchorage of a harness during structural assembly could be detected (Fig. 10), generating a priority alert and automatic report.

Fig. 10. Illustration of a worker improperly anchored to a scaffolded structure and simulated alert issued by the drone (graphic by the author).

The integration of computer vision into safety strategies marks a true paradigm shift in risk management and supervision. It goes beyond expanding the visual scope of safety officers—it equips systems with automated cognitive capabilities to interpret the environment, anticipate hazards, and respond instantly. Unlike traditional inspections—episodic and reliant on human perception—these systems provide continuous, objective surveillance, even in dynamic or hard-to-access environments.

Moreover, their connection to predictive analytics platforms, field devices (drones, wearables), and traceability systems contributes to the development of intelligent safety ecosystems, where prevention becomes an embedded structural component of the digital work environment.

The value of this technology is not limited to detecting obvious risks—it lies in its ability to transform each image into actionable data that supports auditing, training, continuous improvement, and strategic decision-making. In this sense, computer vision becomes a key driver of preventive maturity, offering benefits such as (Mayo de Com, 2021; Torres Abril et al., 2023):

- Continuous, non-intrusive supervision—24/7—with no human fatigue.
- Precise, objective risk detection, eliminating perceptual biases.
- Automation of traditionally manual safety control tasks.
- Generation of auditable evidence to meet internal and external standards.
- Enrichment of predictive models with real-world visual data.

However, its deployment demands careful attention to critical and sensitive issues (Rodríguez-Martínez et al., 2023; Torres Cadillo, 2023):

- Privacy and ethics: Image processing must comply strictly with data protection regulations and avoid disproportionate surveillance practices.
- Contextual model training: Algorithms must be trained with actual images from the operational environment to ensure accuracy and contextual relevance.
- Staff training and acceptance: Workers must understand that the technology is for safety, not punishment—fostering a climate of trust and collaboration.

In conclusion, computer vision applied to occupational safety fundamentally redefines the supervisory role: it doesn't just observe—it interprets and acts. Its real-time understanding of the environment and ability to generate automatic responses make it one of the most transformative developments in contemporary industrial safety.

Far from replacing safety professionals, it enhances their work, expands their perception capacity, and frees up time for high-value tasks like analysis, training, or continuous improvement. Thus, computer vision emerges as a strategic pillar of AI-driven prevention systems.

3.5 Virtual Assistants and Training Chatbots

Conversational artificial intelligence has enabled the development of virtual assistants that are transforming how workers access occupational safety information, receive training, and resolve operational queries. These solutions—powered by natural language processing (NLP) and machine learning technologies—provide constant, accessible, and personalized support tailored to the specific characteristics of each user (Leguizamón Páez & Becerra Correa, 2024).

Key applications include (Goñi Sein, 2024; Nájar Becerra, 2024; Otero Mateo et al., 2023):

- Instant responses to questions about regulations, protocols, and procedures.
- Personalized training based on the worker's profile, work environment, and training history.
- Dynamic knowledge assessments through interactive and adaptive questioning.

Their greatest value lies in their continuous, multichannel availability, which ensures uninterrupted, contextual access to safety knowledge—regardless of time, location, or device.

The digital transformation of occupational risk prevention has fostered new modes of interaction between safety management systems and workers. Virtual assistants and intelligent chatbots have emerged as key tools in this ecosystem, acting as digital extensions of the occupational health and safety (OHS) department (Celi-Parraga et al., 2021; Torres Cadillo, 2023). They not only automate access to information but also deliver training, reinforce key concepts, and provide real-time technical support, all adapted to the user's education level, language, or professional role.

This technology is particularly useful in environments with high staff turnover, multiple job sites, or multilingual teams—where message consistency and training continuity are critical.

A virtual assistant designed for workplace safety typically includes several interlinked modules (Goñi Sein, 2024; Otero Mateo et al., 2023):

- **Conversational interface**: the user's point of contact via text or voice, accessible through corporate apps, browsers, or physical terminals on-site.
- **Semantic engine**: interprets the user's intent, identifies key entities, and understands context. For instance, a query like "Can I use this ladder in an ATEX zone?" receives an accurate, contextualized, and regulation-compliant response.

Fig. 11. Example of a safety-related query highlighting the importance of the semantic engine (graphic by the author).

- Dynamic knowledge base: links to a live repository of technical content (regulations, safety data sheets, manuals, internal protocols), updated in real time.
- Supervised learning system: continuously improves by learning from interactions, correcting mistakes, and fine-tuning responses.

- Personalization engine: customizes content based on the user's profile, language, role, location, and education level, ensuring a relevant and useful experience.

When integrated into training and awareness programs, virtual assistants go beyond answering ad hoc questions. They play an active role in strengthening knowledge and supporting preventive culture. Notable use cases include:

- Personalized microlearning: Short, context-driven modules in text, video, or graphic format. For instance, before entering a chemical hazard area, the assistant may suggest a 90-second video on proper PPE use, followed by a short interactive quiz (Estardid Colom, 2023; Megino Fernández, 2023).
- Dynamic evaluations: Automatically generated adaptive quizzes based on user history or operational indicators. If, for example, recurring issues are detected with lockout/tagout procedures, the system can assign a test to the relevant personnel (Campero-Jurado et al., 2020; Pantoja-Rodríguez et al., 2017).
- Real-time operational support: Instant responses to on-the-job queries about equipment use, applicable regulations, or best practices. For example, a worker may ask if a wooden step ladder is suitable for electrical work at height. The assistant replies with the relevant regulation, an explanatory illustration, and a preventive warning, if necessary (Schor-Landman, 2022).
- Usage traceability and analytics: Every interaction is recorded, enabling the identification of common concerns, assessment of comprehension levels, and fine-tuning of training plans. If multiple workers frequently ask about manual handling, the system can propose a targeted training intervention (Goñi Sein, 2024; Rodríguez Elizalde, 2025).

Integrating virtual assistants into safety systems yields a broad set of strategic benefits:

- 24/7 availability from any device or location.
- Reduced operational workload, freeing safety professionals to focus on high-value tasks.
- Multichannel, multiprofile adaptability, catering to both field workers and technical staff, with content depth tailored to each role.
- Continuous, embedded learning, replacing isolated training sessions with in-context reinforcement.

- Data-driven decision-making, using real user data to improve communication, training, and risk intervention strategies.

Ultimately, these tools contribute to a living safety culture. Rather than replacing formal training programs, virtual assistants complement them with agile, educational, and efficient support. They help transform safety into a continuous, accessible, and worker-centered process—enhancing safety maturity across the organization.

In conclusion, virtual assistants for occupational safety represent a key innovation in the digitalization of workplace risk prevention. By integrating artificial intelligence with pedagogical design, full accessibility, and contextual personalization, they become strategic allies in building safer, better-informed, and more resilient work environments.

3.6 Detection of Fatigue and Emotional States

The early identification of adverse physiological and emotional states—such as fatigue, stress, or emotional dysregulation—represents one of the most relevant emerging applications of artificial intelligence in the field of occupational risk prevention. Thanks to the convergence of biometric sensors, computer vision, and machine learning models, it is now possible to monitor workers' psychophysiological status in real time and anticipate conditions that could compromise their safety or that of the work environment (Galván Guerra et al., 2024; Mayo de Com, 2021).

In particularly demanding environments—such as long shifts, repetitive tasks, work at height, or the handling of heavy machinery—even a slight decline in alertness can result in serious accidents. Artificial intelligence makes it possible to replace subjective and reactive supervision with objective, continuous, and personalized prevention tailored to the worker's real-time condition, enhancing the system's anticipatory capacity (Garimella et al., 2020).

These systems are based on non-invasive technologies that capture, process, and interpret physiological and behavioral signals. Common components include (Megino Fernández, 2023; Nájar Becerra, 2024):

- Biometric sensors embedded in wearable devices (wristbands, vests, or smart helmets) that measure variables such as heart rate, body temperature, heart rate variability (HRV), perspiration, or postural acceleration.

- Computer vision systems capable of analyzing micro facial expressions, blinking patterns, head tilt, or motion trajectories as subtle indicators of physical fatigue or attention loss.
- Voice and speech analysis in conversational assistants, examining variables such as tone, cadence, speech speed, or atypical pauses as potential signs of stress or cognitive fatigue.
- Machine learning models that integrate these variables to build individual risk profiles, generate preventive alerts, and activate real-time tailored recommendations.

These technologies are being piloted or implemented in specific labor contexts. In sectors such as construction, mining, logistics, or the chemical industry, pilot experiences have shown promising results. Notable use cases include (Macías García, 2022; Rojas-Gualdron, 2022):

- Prevention of accidents due to accumulated fatigue: By detecting physiological patterns consistent with physical or cognitive exhaustion, the system can recommend strategic breaks, reschedule shifts, or activate task rotations. For example, if a worker during an intensive shift exhibits a sustained drop in heart rate variability along with postural signs of inattentiveness, the system will issue an alert to interrupt the activity and restore optimal activation levels (Fig. 12).
- Reducing risks associated with stress or altered emotional states: Continuous monitoring can identify signs of high anxiety, emotional blockage, or impulsive reactions. In a critical operation, a technician whose voice suddenly shows changes in rhythm or intonation may be flagged by the system as emotionally compromised, triggering an immediate replacement or additional evaluation (Fig. 13).

Fig. 12. Representation of a worker in a fatigue alert situation after a drop in heart rate variability (graphic by the author).

Fig. 13. Representation of a technician whose vocal modulation is interpreted by the system as indicative of a possible emotional alteration, activating adaptive preventive measures (graphic by the author).

- Dynamic task allocation according to psychophysiological state: Individualized monitoring enables reconfiguration of work planning based on the real workload borne by each worker. For instance, if several night shift workers show signs of visual fatigue, the system may reassign less visually demanding tasks or extend rest periods between inspection cycles (Fig. 14).

Fig. 14. Representation of several night shift workers with indicators of visual fatigue, prompting the system to adapt the planning by redistributing less visually demanding tasks and extending rest intervals (graphic by the author).

From a global perspective, the integration of smart systems for detecting fatigue and emotional states represents a major step toward person-centered prevention. Key benefits include (Bonfante et al., 2021; Constance Obiuto et al., 2024; Rojas-Gualdron, 2022):

- Continuous and contextual monitoring of psychophysiological status incorporating human factors traditionally excluded from preventive analysis.
- Objective and early identification of risk conditions, preventing them from materializing into incidents or operational errors.

- Optimization of workload through preventive adjustments in pacing, shift scheduling, or task allocation aligned with the worker's actual functional capacity.
- Reduction of human errors derived from fatigue or cognitive overload, particularly in critical tasks where sustained attention and rapid decision-making are essential.

Additionally, integration with other digital platforms in risk management enables a more comprehensive, interconnected, and human-sensitive risk analysis model, promoting proactive and predictive prevention.

Implementation of these technologies must be approached with responsibility, transparency, and respect for individual rights. It is crucial to define clear boundaries regarding (Campero-Jurado et al., 2020):

- Monitoring purposes: it must focus on prevention and worker well-being, avoiding punitive or intrusive use.
- Data privacy and protection: biometric and emotional data processing must strictly comply with data protection regulations and require informed consent from workers.
- System transparency and governance: data collection, analysis, and interpretation mechanisms must be auditable, understandable, and accessible to all stakeholders.

Moreover, these tools should not foster dynamics of distrust or stigmatization. Their value lies in being perceived as instruments for care and sustainability of human effort—not as covert surveillance mechanisms.

Ultimately, we are witnessing a shift toward a new paradigm of prevention based on internal states. The adoption of fatigue and emotional detection technologies ushers in a new stage in the evolution of occupational safety. This approach moves away from a purely normative and behavioral model toward one rooted in the real-time condition of the worker, with adaptive and immediate responsiveness. It means acting before risk manifests, addressing its physiological and emotional precursors.

Thus, the most advanced prevention recognizes that monitoring the environment is insufficient without integrating the internal state of the person. In this new paradigm, the human factor is no longer a source of uncertainty—it becomes the core of the preventive system.

3.7 Automation of Audits and Regulatory Compliance

Managing regulatory compliance in occupational risk prevention requires continuous supervision of workplace conditions, equipment, procedures, and documentation mandated by current legislation. Traditionally carried out manually and periodically, this process entails significant operational workload and is prone to human error, delays, and a reactive focus based on post-factum findings. The integration of artificial intelligence and automation technologies transforms this model into one of continuous, autonomous auditing, grounded in verifiable digital evidence (Chicaiza Ortiz, 2024).

By combining computer vision, document verification algorithms, connected sensors (IoT), and intelligent analytics platforms, it is now possible to develop a far more efficient, traceable, and proactive auditing system. Such a system is capable of detecting non-compliance in real time, triggering automated alerts, documenting deviations, and linking them to supervised corrective action plans. This evolution enhances the maturity of preventive systems by embedding compliance directly into everyday operational workflows—rather than treating it as a discrete administrative milestone.

The automation of compliance is built on a modular architecture that typically includes the following components:

- Automated data capture (structured and unstructured): Sensors, cameras, wearables, digital forms, and SCADA systems continuously feed data on facility conditions, worker activity, and environmental parameters.
- Regulatory engine and digital rule matrix: Incoming data is cross-referenced with encoded regulatory requirements, including provisions from Spain's Law 31/1995 on Occupational Risk Prevention, Royal Decree 1215/1997 on the safe use of work equipment, and standards such as ISO 45001. Artificial intelligence automatically identifies deviations from legal obligations.
- Anomaly detection and alert generation: Algorithms detect irregular conditions, such as missing personal protective equipment (PPE), inadequate signage, expired medical clearances, or misuse of equipment.
- Audited recordkeeping and automated report generation: All actions and events are timestamped, documented with visual evidence and metadata, ensuring full traceability for both internal and external audits.

- Dashboards and predictive analytics: Data are integrated into interactive dashboards that help prioritize field inspections, analyze compliance trends, and forecast potential risk areas.

In terms of practical applications, some of the most widely adopted use cases in occupational safety management include:

- Automated verification of safety checklists at site entrances, for equipment and high-risk zones via computer vision, ensuring real-time compliance with baseline safety requirements.
- Smart document control, where AI-based systems verify the validity of certificates, permits, safety data sheets, or training records, and generate alerts for upcoming expirations.
- Visual detection of physical non-compliances, such as improperly installed safeguards, obstructed access points, or defective/missing signage—even in areas that are not routinely inspected.
- Digital tracking of corrective actions, linking each non-conformance to a specific resolution task, responsible person, and deadline, with alerts for overdue or neglected actions.
- Remote auditing and AI-based validation, especially useful in dispersed worksites or hard-to-reach areas, where field staff can submit photos or videos for automated assessment and validation of safety conditions or task closures.

When analyzing operational and strategic benefits, it is clear that automation of regulatory compliance in occupational risk prevention provides:

- A drastic reduction in manual workload, freeing up safety professionals for higher-value activities.
- Improved objectivity and consistency in the identification of non-compliances.
- Immediate response to critical deviations through intelligent alerts.
- Structured, auditable evidence generation—vital for legal inspections and certification audits.
- Enhanced preventive analysis, by embedding regulatory logic within the broader digital safety management system.

This model does not replace the safety professional—it empowers them. It transforms their role from operational executor to strategic leader, focused on analysis, continuous improvement, and fostering a robust safety culture.

Under this paradigm, auditing evolves from a static, reactive exercise into a continuous, integrated, and anticipatory function. The intelligent system does not wait for failure to occur: it detects it, documents it, and ties it to its solution before it escalates into legal, financial, or human consequences. This represents a qualitative leap in risk management—where technological intelligence becomes the guarantor of legal compliance and a strategic ally to the occupational safety practitioner.

3.8 Integration with Other Technologies (IoT, Big Data, BIM)

Artificial intelligence applied to occupational risk prevention is not an isolated solution; rather, it reaches its full potential when harmoniously integrated with other emerging digital technologies. This technological convergence gives rise to an interconnected, context-aware preventive ecosystem, capable of dynamically adapting to the operational variability of modern work environments (Corvalán, 2018).

Among the technologies that—when combined with artificial intelligence—greatly enhance its impact, three foundational pillars stand out: the Internet of Things (IoT), Big Data, and Building Information Modeling (BIM) (Chernyshev et al., 2022; Mazzetto, 2024; Morales-Conejo, 2021). Each contributes a complementary layer of information and analysis which, when integrated, transforms safety management into a more precise, proactive, and data-driven discipline:

- **Internet of Things (IoT): Intelligent environmental sensing.** The deployment of distributed sensors throughout the work environment enables real-time data collection on ambient conditions, machinery performance, PPE status, personnel movement, or the presence of hazardous agents such as gases, noise, or vibrations. This constant sensing stream provides artificial intelligence with granular and contextualized data, allowing the construction of dynamic risk models and the activation of autonomous safety protocols.

  For example, in a confined space, a sensor detects a critical concentration of ammonia. The AI system assesses the severity of the hazard, triggers a localized alarm (Fig. 15), and may even initiate automatic access lockdown through presence control systems.

Fig. 15. Example of an ammonia detection sensor integrated into a high-risk workplace (graphic by the author).

- Big Data: Pattern recognition and emerging risk prediction. The large-scale processing of complex variables—both structured and unstructured, historical and real-time—enables AI systems to identify correlations, trends, and cause-effect relationships that would otherwise go unnoticed. This approach not only improves responsiveness to individual events but also enables anticipatory prevention based on the detection of behavioral patterns and risk profiles.

  For instance, after analyzing millions of incident records, AI can determine that failures in preventive maintenance significantly increase the likelihood of entrapment accidents. As a result, the system recommends strengthening specific controls at critical points—before the risk materializes.

- BIM (Building Information Modeling): Proactive simulation from the design phase. Integrating AI with BIM platforms introduces a crucial prospective capability. Through three-dimensional models of construction sites and facilities, it becomes possible to simulate working environments prior to execution, anticipate hazards, and assess the impact of various safety measures during the early stages of a project.

A representative example would be a project involving work at height, where the BIM model identifies critical proximity between working platforms and suspended loads (Fig. 16). The AI system evaluates multiple interaction scenarios and proposes design reconfigurations to reduce the risk of falling objects.

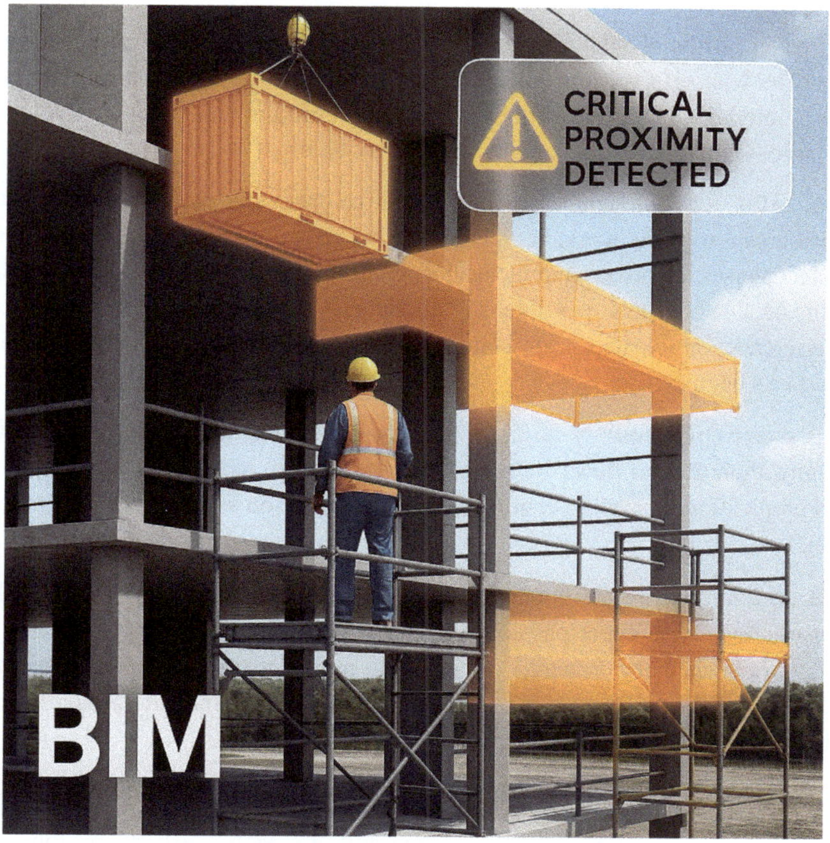

Fig. 16. Example of a BIM model detecting interference between working platforms and suspended loads (graphic by the author).

The simultaneous integration of artificial intelligence with enabling technologies such as IoT, Big Data, and BIM is profoundly redefining the role of risk prevention in complex work environments. This new paradigm establishes an intelligent, systemic, and continuously evolving system that surpasses traditional frameworks and places real-time, contextualized data at the center of the prevention process.

The convergence of these technologies allows for permanent monitoring of environmental, structural, organizational, and human conditions, drawing from heterogeneous data sources (physical sensors, digital records, human variables, and simulations) that are continuously updated and cross-referenced in real time. This multisensory monitoring enables early risk detection, drastically reducing response time to deviations or emerging threats.

Moreover, artificial intelligence adds a contextual analytical layer: it not only identifies what is happening, but infers why it is happening and what could happen if no intervention occurs. This transforms prevention into a predictive—not reactive—function capable of anticipating risk before it occurs.

The BIM environment provides a three-dimensional, dynamic virtual representation of the worksite, integrating not only structural elements but also project phases, involved equipment, and potential hazards. Through its integration with AI, it becomes possible to simulate operational scenarios, plan for interference, anticipate unsafe conditions, and test preventive strategies before real tasks begin.

This prospective capacity reshapes prevention at the design level, enabling the optimization of workflows, definition of critical zones, adjustment of schedules, and preparation of specific contingencies—based not on assumptions, but on real data.

The fusion of artificial intelligence, IoT, Big Data, and BIM enables the construction of a cyber-physical, adaptive architecture in which preventive resources—whether human, technical, or logistical—are dynamically reconfigured based on the simultaneous analysis of thousands of variables. This allows organizations to:

- Prioritize interventions where the risk is real, not just regulatory.
- Personalize safety protocols according to each worker's profile and contextual factors.
- Optimize allocation of resources (supervision, signage, collective protections) in real time.

This new approach turns prevention into a distributed, responsive, and self-regulating system in which technology enhances, rather than replaces, the occupational safety professional—elevating their role to that of a strategic decision-maker supported by evidence.

This is no longer about detecting risk once it occurs. It is about predicting it before it arises, simulating its emergence, and mitigating its impact before it becomes harm. Data becomes the shared language between machines, systems, and people—while safety evolves into a continuous, proactive, and knowledge-centered process.

3.9 Operational Guide for the Implementation of AI in Occupational Risk Prevention

The adoption of artificial intelligence in occupational risk prevention should not be approached as a mere technological upgrade. Rather, it entails a profound transformation of preventive culture, organizational architecture, and decision-making processes. Proper integration demands technical rigor, ethical commitment, and a clear vision of the value it brings: anticipating risks, protecting lives, and strengthening the professional role of safety technicians (Schor-Landman, 2022).

Moreover, when deployed alongside enabling technologies such as the Internet of Things (IoT), Big Data, and Building Information Modeling (BIM), AI redefines prevention as a dynamic, intelligent, and proactive function, capable of operating in real time in complex environments based on continuous, contextual, and predictive information (Chernyshev et al., 2022; Mazzetto, 2024; Morales-Conejo, 2021).

This new paradigm transforms workplace safety into a self-regulating cyber-physical system, where data not only reflect what is happening but help to understand why it happens, anticipate risk scenarios before they materialize, and optimize preventive decisions from the design phase onward (Goñi Sein, 2024; Macías García, 2022; Rodríguez Elizalde, 2025).

What follows is a structured roadmap for the responsible, strategic, and effective implementation of AI systems in occupational risk prevention.

3.9.1 Preliminary Assessment

The first step is to conduct a thorough diagnosis of the organization's digital maturity and preventive capacity. This analysis should consider (Chernyshev et al., 2022; Mazzetto, 2024):

- The existing level of sensorization (IoT infrastructure, SCADA systems, digitalized OSH platforms).
- The availability and quality of historical data on accidents, near misses, safety observations, inspections, and training.
- Identification of processes that can benefit from AI, such as PPE compliance monitoring, environmental condition analysis, or fatigue tracking.
- A review of the applicable regulatory framework, with particular attention to data protection laws, electronic surveillance, and the principles of the European AI Act.

This assessment helps avoid unnecessary or ineffective solutions and allows for the prioritization of high-impact and realistic use cases.

3.9.2 Definition of Objectives and Use Cases

Any initiative should be driven by specific, measurable safety objectives aligned with the core principles of prevention (see Section 2.2). High-impact use cases include:

- Automatic detection of PPE compliance through computer vision.
- Predictive accident analysis based on historical and contextual patterns.
- Continuous monitoring of physiological or environmental variables (temperature, CO_2, vibrations) via connected sensors.

Each use case should be tied to key performance indicators (KPIs) such as audit time reduction, regulatory compliance improvements, or incident reduction in high-risk areas.

3.9.3 Tool and Vendor Selection

Technology selection should be based on technical, ethical, and operational criteria—not just commercial ones. It is essential that AI tools (Chicaiza Ortiz, 2024):

- Are transparent and understandable for safety teams (avoiding “black box” algorithms).
- Offer traceability and explainability of automated decisions.

- Are designed with a privacy-by-design approach and comply with the General Data Protection Regulation (GDPR).
- Are interoperable with existing systems (ERP, construction platforms, BIM, etc.).

Vendor experience in real-world work environments, ethical integrity, and post-implementation support capacity should also be evaluated.

3.9.4 Personnel Training and Operational Integration

Technology alone does not generate impact without integration into real workflows. This phase includes (Estardid Colom, 2023; Megino Fernández, 2023):

- Targeted technical training for OSH professionals on AI fundamentals, system validation, and bias supervision.
- Design of human-in-the-loop protocols to ensure critical decisions remain supervised by domain experts.
- Promotion of interdisciplinary collaboration between safety technicians and data engineers to interpret information from both technical and contextual perspectives.

This collaborative approach is essential to avoid design flaws, technical misunderstandings, or algorithmic overdependence.

3.9.5 Continuous Evaluation and Adaptive Improvement

AI implementation should be an iterative process, grounded in continuous improvement cycles (Dolhopolov et al., 2023; Rojas-Gualdron, 2022). Recommended practices include:

- Regular review of KPIs and their evolution over time.
- Systematic feedback collection from end users (workers, safety personnel, supervisors).
- Technical and ethical audits to identify degradation, reliability losses, or unintended side effects (e.g., reduced professional judgment or overfitting).
- Model updates, retraining with new data, and redesigns when operating conditions change.

Only through an active improvement culture can AI evolve alongside the reality it aims to protect, rather than becoming obsolete or disconnected from the working environment.

Implementing AI in occupational safety is not a matter of buying technology—it is about embracing a new way of thinking about workplace health and safety. An intelligent system does not replace responsibility or professional judgment; it enhances them with new analytical, adaptive, and predictive capabilities.

But for this potential to be realized, it must be implemented with rigor, insight, and a clear ethical orientation (Lassi, 2023).

4 Advantages of Artificial Intelligence in Occupational Risk Prevention

4.1 Enhanced Preventive Capacity

The incorporation of artificial intelligence (AI) into occupational risk prevention is not merely a technological innovation; it represents a qualitative leap in the system's ability to anticipate, adapt, and act with greater precision and efficiency. Unlike traditional approaches—centered on identifying hazards only after they have caused harm or on static regulatory compliance—AI enables the development of dynamic prevention models, grounded in real and contextualized data, which broaden both the reach and impact of preventive actions (Goñi Sein, 2024; Macías García, 2023).

One of AI's greatest strengths lies in its capacity to process vast volumes of heterogeneous data—both structured and unstructured—in real time and from multiple sources. This analytical power enables the detection of subtle correlations, emerging patterns, and incipient risk situations that would go unnoticed with conventional methods. As a result, the system not only reacts to danger but can anticipate it and activate responses before any tangible manifestation occurs (Rodríguez Elizalde, 2025).

Moreover, AI allows for prevention to be adapted to each specific situation, personalizing protocols, measures, and training content according to variables such as the worker's profile, the task being performed, their physiological condition, environmental factors, or the type of operational environment. This contextual adjustment replaces generic solutions with more accurate, relevant, and effective interventions.

Additionally, the automation of tasks such as visual supervision, document control, exposure tracking, or the issuance of alerts reduces the operational burden on human systems, freeing up resources that can be redirected to higher-value tasks (Massiris et al., 2021). In this way, preventive coverage is continuously expanded without proportionally increasing the size of the workforce.

The use of predictive models and machine learning algorithms also transforms risk assessment into an evolving process. Systems learn and improve as they analyze new situations, enhancing their ability to anticipate hazardous conditions and propose corrective actions based on accumulated empirical evidence.

In summary, AI does not replace prevention professionals—it multiplies their operational capacity, equipping them with tools that expand their field of observation, accelerate their responsiveness, and deepen their analysis. The result is a preventive system that is more anticipatory, personalized, and resilient—ready to address the challenges of increasingly dynamic, technological, and complex work environments.

4.2 Increased Effectiveness and Speed in Decision-Making

One of the most significant impacts of artificial intelligence in the field of occupational risk prevention is the substantial improvement in the quality, speed, and timeliness of decision-making. In work environments where risk levels can shift rapidly—due to changes in physical conditions, task organization, or worker status—having systems that can analyze data in real time and trigger immediate responses can mean the difference between preventing an incident and merely reacting after it occurs.

Traditionally, preventive decisions have been based on periodic inspections, post-hoc reports, or subjective evaluations. While useful, these methods are clearly limited in terms of immediacy, accuracy, and adaptability. AI overcomes these limitations by integrating data from multiple sources—environmental sensors, cameras, historical records, biometric data, or weather conditions—and processing it using algorithms that identify complex patterns, detect deviations, and generate recommendations or alerts automatically and at the right time (Galván Guerra et al., 2024).

This automated analysis not only speeds up decision-making but grounds it in objective evidence, reducing uncertainty and the likelihood of human error. For instance, an AI system can recognize that a combination of critical tasks, adverse weather conditions, and high workload constitutes a high-risk scenario, and can immediately activate corrective measures—without waiting for manual intervention.

Furthermore, predictive models go beyond responding to current conditions—they anticipate likely scenarios, allowing for advanced planning, resource reallocation, and exposure minimization (Zamora-Cevallos et al., 2024). The decision-making process thus shifts from being reactive to proactive and strategic.

Another key benefit is the reduction in response times during critical situations. Automated systems can, for example, block access, issue alarms, or halt operations in the presence of hazardous conditions—without relying on the immediate availability of supervisory staff. This capacity for autonomous action, based on validated technical criteria, enables decisive seconds to be gained, which may prevent an accident or mitigate its severity.

Altogether, AI provides the prevention system with superior operational agility and informed decision-making capabilities compared to conventional models. Decisions are not only made faster, but also with greater accuracy and contextual depth, thanks to real-time, multidimensional analysis. This evolution fosters a more effective, adaptive, and resilient prevention strategy—aligned with the complexity and dynamism of today's work environments.

4.3 Continuous and Real-Time Monitoring

One of the most transformative pillars of artificial intelligence applied to occupational risk prevention is the implementation of continuous and real-time monitoring systems. These systems are capable of supervising the environment, workers' physical condition, and equipment behavior with uninterrupted accuracy from multiple simultaneous vantage points. This model definitively breaks with the episodic and reactive nature of traditional prevention, ushering in a permanent, adaptive, and autonomous safety framework—even in highly variable and complex operational settings.

As previously discussed in sections 3.2 Real-Time Monitoring, 3.4 Computer Vision for Risk Detection, and 3.5 Virtual Assistants and Training Chatbots, this capability relies on an integrated architecture combining IoT sensors, computer vision, biometric analysis, and distributed computing (in the cloud or at the edge). This infrastructure generates a continuous flow of data—including temperature, gas levels, vibrations, presence detection, use of personal protective equipment (PPE), posture analysis, and signs of fatigue—that is interpreted in real time by AI algorithms capable of detecting anomalies, anticipating risks, and activating automated responses without human intervention (Garimella et al., 2020).

Unlike models based on periodic inspections or manual logs, continuous monitoring captures critical events at the exact moment they occur—even in the absence of supervisors. Thus, safety becomes a structural function of the digital environment, ensuring constant and objective coverage.

Furthermore, permanent surveillance enhances both granularity and traceability of preventive controls. Every collected data point is timestamped and contextualized, enabling detailed incident reconstruction, more rigorous root cause analysis, and evidence-based decision-making. This traceability also strengthens regulatory compliance and prepares the organization more effectively for internal and external audits.

From a strategic perspective, real-time monitoring goes beyond reactive capability, feeding into predictive systems and continuous improvement models. The accumulation of data allows for the identification of recurring patterns, assessment of intervention effectiveness, and dynamic adjustment of preventive resources. In this sense, the system does not merely observe—it learns, adapts, and improves its performance over time.

Another key advantage is the expansion of operational reach. These systems make it possible to monitor remote areas, hard-to-reach locations, or scenarios where human oversight is limited—such as vertical construction sites, night shifts, or high-risk industrial facilities—ensuring more robust and equitable prevention.

Overall, continuous and real-time monitoring redefines prevention as a connected, intelligent, and evolving process. It is no longer just about seeing more—but about understanding better and acting faster, thanks to a technological infrastructure that positions safety as a core attribute of the production system.

4.4 Personalization of Prevention

Artificial intelligence has opened the door to a new approach in occupational risk prevention: data-driven personalized prevention. Unlike conventional models, which apply homogeneous criteria to broad groups of workers, AI enables the design and implementation of preventive strategies tailored to the individual profile, the specific conditions of the job, and the real dynamics of the work environment.

This paradigm shift addresses a structural need within the preventive system: acknowledging that risk does not impact everyone in the same way, even in seemingly similar contexts. Factors such as experience, physical condition, mental workload, cumulative exposure, work schedules, or language environment directly affect each individual's level of vulnerability.

Thanks to technologies such as machine learning, computer vision, and natural language processing, it is now possible to accurately capture and analyze individual data, correlating it to build dynamic risk profiles. Based on these, intelligent systems can recommend personalized measures: from safer task assignments and strategic breaks, to ergonomic adjustments or targeted training reinforcement (L. D. Jones et al., 2018).

Personalization also extends to training. Content can be adapted to a worker's knowledge level, language, incident history, or learning style, improving training effectiveness and its connection with operational reality.

As an illustration, Table 7 presents concrete examples of personalized prevention based on worker profiles:

Table 7. Illustrative examples of individualized preventive adjustments according to worker profile (table by the author).

Worker Profile	Personalized Preventive Measure
Operator performing physically demanding tasks	Task rotation, active breaks, and specific ergonomic recommendations
New worker performing critical tasks	Assignment of a safety mentor, enhanced initial training, and increased supervision
Employee with previous fatigue or stress episodes	Continuous biometric monitoring and recommendations for breaks or temporary task reassignment
Personnel exposed to hazardous substances	Enhanced environmental monitoring, customized PPE, and more frequent medical check-ups
Worker in a multilingual environment	Preventive training in native language and real-time virtual assistance
Technician working extended night shifts	Adjustments to lighting and work pace, along with monitoring of rest cycles

Additionally, personalization allows the deployment of dynamic protocols that adjust in real time. For instance, if a system detects signs of cognitive fatigue in a worker about to operate critical machinery, it can recommend temporary reassignment or immediately reinforce supervision—without waiting for the risk to materialize.

This approach not only increases preventive effectiveness but also fosters a person-centered safety culture that acknowledges the uniqueness of each worker and abandons one-size-fits-all logic. Personalized prevention does not mean arbitrary segmentation; it means designing fairer, more efficient, and more contextualized solutions.

Moreover, this capacity for individual adjustment enables the preventive system to remain effective even in changing work environments, with shifting teams or non-standard conditions. AI provides prevention with unprecedented agility, offering granular, adaptive control that enhances anticipation, precision, and the impact of each intervention.

However, this level of personalization also raises important ethical challenges (Chicaiza Ortiz, 2024; Lassi, 2023; Terrones Rodríguez, 2018). The collection and analysis of sensitive data—such as a worker's physical, emotional, or behavioral state—require responsible and transparent management. Continuous monitoring and algorithmic classification of risk profiles must be handled with particular attention to privacy, informed consent, and data protection, especially when they directly affect working conditions.

These implications are not merely technical—they affect the legitimacy and social acceptance of the system. Therefore, they will be thoroughly analyzed in the next chapter, dedicated to the risks, limitations, and ethical implications of using AI in prevention.

4.5 More Effective Training

Training is one of the fundamental pillars of occupational risk prevention, yet it is also one of the most vulnerable areas to inefficiency when approached uniformly, without context, or disconnected from operational reality. The incorporation of artificial intelligence into training processes allows these limitations to be overcome, endowing preventive education with greater relevance, adaptability, follow-up capacity, and real impact.

Through machine learning algorithms and natural language processing, AI makes it possible to design personalized training paths tailored to each worker's profile, based on variables such as the type of task, experience, incident history, results of previous evaluations, or learning preferences (Freré Arauz et al., 2022; Megino Fernández, 2023; Sanz-Valero & Martínez-Aguirre, 2024). This adaptive approach ensures that each person receives the content they need, at the right time, and in the most effective format for their learning. This was previously discussed in sections 3.5 Virtual Assistants and Training Chatbots and 3.3 Predictive Accident Analysis.

Moreover, intelligent training systems allow real-time evaluation of content comprehension and adapt the training progression based on individual performance. AI is capable of identifying error patterns, knowledge gaps, and proposing automatic reinforcement, avoiding both information overload and superficial learning.

Another key advantage is the ability to integrate training into the operational workflow, through micro-content accessible from mobile devices, control terminals, wearables, or conversational assistants. Thus, training becomes a continuous, contextual, and action-oriented process rather than an isolated and occasional activity.

In complex environments or those with high staff turnover, AI also facilitates automatic translation of content, adaptation to various levels of digital literacy, and the generation of specific materials for critical tasks or those with a low margin for human error.

The traceability provided by these systems allows occupational safety and health professionals to precisely monitor both individual and collective progress, evaluating not only the degree of training compliance but also knowledge retention and the actual effectiveness of the actions carried out (Goñi Sein, 2024; Macías García, 2022).

Far from replacing human-led training, AI complements, guides, and reinforces it with a level of adaptation and analysis that traditional methods cannot reach. The result is a learner-centered training model, closely aligned with real risks and far more effective from a preventive perspective.

The following table summarizes the key differences between traditional preventive training and AI-assisted training:

Table 8. Comparative table between traditional training and AI-assisted training (table by the author).

Aspect	Traditional Training	AI-Assisted Training
Approach	Uniform and generalist	Contextual and person-centered
Personalization	Limited or non-existent	Highly personalized by profile and performance
Delivery format	In-person sessions or static materials	Multi-format, interactive, and field-accessible
Learning assessment	Standardized periodic tests	Continuous and dynamic evaluation
Adaptation to performance	Not adjustable to worker's progress	Automatic adjustment to errors and needs
Accessibility	Dependent on classrooms or scheduled moments	Ubiquitous, multi-platform, and real-time
Traceability	Manual and limited record-keeping	Complete and automated monitoring
Job relevance	Disconnected from real operational context	Integrated into real workflow and risks

This new approach not only improves knowledge acquisition and retention but also helps consolidate a more active, participatory safety culture aligned with the real needs of the productive environment. Thus, training ceases to be merely a bureaucratic requirement and becomes a strategic resource for anticipation, intervention, and continuous learning.

4.6 Reduction of Human Error

In occupational risk prevention systems, human error remains one of the primary triggers of incidents—particularly in complex operational environments, under time pressure, or in contexts with high cognitive load. Artificial intelligence does not aim to eliminate human intervention, but rather to minimize the margin of error through technologies that offer assistance, supervision, and real-time responsiveness (Schor-Landman, 2022; Vélez Aspiazu et al., 2024).

The main advantage of AI in this field lies in its ability to detect error-prone conditions before the error actually occurs. By analyzing contextual, behavioral, and physiological data—such as fatigue, distraction, sustained stress, or inadequate operating patterns—machine learning algorithms can anticipate critical deviations and trigger preventive corrective actions.

Moreover, intelligent systems can act as operational assistants, guiding workers through complex tasks. Through intuitive interfaces, step-by-step instructions, or automated alerts, AI reduces ambiguity, simplifies procedures, and validates critical decisions—acting as a second layer of control that reinforces reliability without undermining operator autonomy.

This technological support also allows for continuous monitoring of behavior, capturing micro-events that may seem harmless individually but can accumulate and lead to more serious incidents. This constant oversight proves particularly valuable in high-risk activities or those with low tolerance for error, such as working at heights, handling hazardous substances, or operating heavy machinery.

On another level, the value of AI goes beyond direct prevention. Its traceability and post-incident analysis capabilities allow for the accurate reconstruction of circumstances leading up to an error, identification of root causes, and the redesign of protocols based on objective and verifiable data.

This approach highlights a key principle: AI does not replace the worker—it protects and empowers them. These systems are designed to enhance human judgment, not eliminate it (Fernández-Costales Muñiz, 2020; Schor-Landman, 2022). In fact, rather than displacing jobs, AI tends to redefine professional roles, freeing workers from routine tasks or constant surveillance and enabling them to focus on data analysis, strategic planning, and continuous improvement. This reality challenges the simplistic narrative that associates AI with indiscriminate job destruction.

From a strategic perspective, it is useful to examine how different types of human error can be addressed through AI-based solutions. Table 9 summarizes some of the most relevant correspondences:

Table 9. Human Errors and AI-Based Solutions (table by the author).

Type of Human Error	AI-Based Intervention
Accumulated fatigue	Continuous biometric monitoring and activation of adaptive break recommendations
Distraction or attention lapses	Computer vision and real-time alerts for unsafe behavior
Lack of training or knowledge	Personalized training and context-aware virtual assistance
Time pressure or excessive multitasking	Dynamic task prioritization and assistants that modulate operational workload
Misinterpretation of instructions	Step-by-step guided instructions and action validation through intelligent assistants
Repetition of errors or ingrained bad habits	Behavioral pattern analysis and automated corrective feedback

In conclusion, reducing human error through artificial intelligence does not mean replacing professional judgment, but rather enhancing its reliability and mitigating its inherent vulnerabilities. This approach helps to build a more robust and mature prevention system—one in which risk is managed through anticipation, continuous learning, and intelligent collaboration between people and technology.

4.7 Resource Optimization

One of the most tangible benefits of artificial intelligence (AI) applied to occupational risk prevention is its capacity to optimize the use of preventive resources, both from a technical and economic-organizational perspective. Unlike traditional models based on rigid planning and uniform resource distribution, AI introduces a dynamic, context-aware, data-driven approach that enables doing more with the same resources—or even with fewer.

First, AI contributes to a significant reduction in costs associated with workplace accidents (Toro Lazo, 2024; Tuya Gonzalez et al., 2011), by decreasing the frequency and severity of incidents through enhanced anticipation, more precise decision-making, and earlier intervention. Fewer incidents mean fewer sick leaves, legal disputes, operational disruptions, or repair expenses—directly impacting the profitability of the productive system (Fernández-Costales Muñiz, 2020; Macías García, 2022).

Second, AI is crucial for minimizing unplanned downtime, thanks to its ability to detect deviations in real time, predict failures, and trigger automatic responses. This is particularly critical in sectors such as construction, logistics, or heavy industry, where even short interruptions can cause cascading effects (A. Jones, 2024).

Moreover, AI allows for prioritizing preventive investments based on objective criteria such as exposure levels, incident history, task criticality, or current environmental conditions. Resources are thus allocated where they are most needed—not where organizational inertia dictates.

A particularly relevant application is the efficient management of specialized human capital in occupational risk prevention. By automating routine tasks such as document verification, visual supervision, or the tracking of corrective actions, technical professionals are freed to focus on higher-value activities: data analysis, preventive planning, strategic training, or continuous improvement.

Table 10 provides concrete examples of how different preventive resources can be optimized through the use of AI:

Table 10. Examples of Resource Optimization through AI (table by the author).

Preventive Action Resource	AI-Based Optimization
Safety inspections	Dynamic prioritization of high-risk zones for targeted inspections
Worker training	Content adapted by user profile, performance, and context
Use of PPE	Distribution based on tasks, specific risks, and environmental conditions
Assignment of technical staff	Efficient deployment based on alerts, risk indicators, or local needs
Tracking of corrective actions	Automated compliance monitoring and pending task notifications
Preventive maintenance planning	Early anomaly detection and predictive scheduling based on real-time data

This intelligent optimization does not imply budget cuts or resource downgrading, but rather a strategic, efficient, and responsive approach tailored to real working conditions. AI does not eliminate resources—it redistributes them more effectively, expands their reach, and enhances their impact, aligning efforts with areas of highest preventive need.

Overall, AI enables the configuration of a more efficient, results-oriented management system, grounded in empirical evidence. It reduces costs without compromising safety, strengthens organizational resilience, and ensures a proportional, flexible, and adaptive allocation of available resources.

4.8 Strengthening the Preventive Safety Culture

Beyond its technical, operational, or economic benefits, one of the most profound contributions of artificial intelligence (AI) to occupational risk prevention is its ability to foster a more mature, participatory, and sustainable safety culture. The introduction of intelligent technologies does not merely automate functions—it drives a structural transformation in how prevention is perceived, understood, and experienced within organizations.

Traditionally, safety culture has been promoted through awareness campaigns, periodic training, and direct supervision. While useful, these approaches face limitations in terms of personalization, continuity, and connection to workers' lived experiences. AI adds a new dimension: a culture driven by data, informed participation, and constant feedback.

A key element is AI's ability to make risk visible in real time. Automated alerts, contextual visualizations, and on-demand assistance systems help every worker understand how their decisions and habits impact collective safety. When a worker receives a notification for not wearing personal protective equipment (PPE), or sees data showing reduced incidents thanks to safe behavior, it strengthens the link between individual actions and collective outcomes.

Moreover, intelligent systems promote real and equitable inclusion in access to preventive resources. Personalized content, automatic translation, accessibility across multiple devices, and contextualized support ensure that any worker—regardless of language, educational background, or job role—can actively participate in the safety system. This aligns with the principles established in Article 11 of Directive 89/391/EEC, and Articles 18, 33, 34, and 35 of Spanish Law 31/1995 on Occupational Risk Prevention. This democratization of knowledge is essential to building a shared culture of safety.

Additionally, AI-powered traceability enables the identification of cultural resistance, entrenched risky practices, or inefficient organizational dynamics. These insights not only help correct deviations but also allow for deep redesigns of prevention strategies, addressing root causes rather than mere symptoms.

AI also facilitates the recognition and reinforcement of positive behaviors through personalized feedback systems, individual dashboards, and performance-based safety indicators. This approach boosts personal commitment, a sense of ownership, and motivation to maintain proactive, responsible attitudes toward risk.

In short, a strong safety culture cannot be imposed—it must be built through understanding, participation, and trust in the system. AI serves as a facilitator by providing tools that help workers better understand risk, make more informed decisions, and reinforce safe behaviors as part of daily practice.

More than a technological revolution, AI enables a cultural evolution: a shift from reactive, compliance-driven models to a living, distributed, and knowledge-based safety culture, where prevention is not enforced—but embedded.

5 Risks, Limitations, and Ethical Implications of Using Artificial Intelligence in Occupational Risk Prevention

The integration of artificial intelligence into occupational risk prevention systems opens up an unprecedented landscape of opportunities: greater anticipation capacity, personalized interventions, faster decision-making, and a substantial improvement in operational efficiency. However, this technological transformation is not without its risks, limitations, and ethical implications, which must be addressed with rigor and responsibility.

Failing to consider these aspects could compromise not only the technical reliability of the system, but also its social acceptance, legal legitimacy, and organizational sustainability. Artificial intelligence, far from being a neutral tool, reflects and amplifies the human decisions behind its configuration—from algorithm design and data selection to governance and usage criteria.

The purpose of this chapter is to critically and constructively examine the main concerns associated with the use of artificial intelligence in the field of occupational risk prevention. This analysis does not aim to hinder development, but rather to provide a more solid, fair, and transparent foundation. Only by understanding its weaknesses can we build safer, more inclusive, auditable, and ethically responsible preventive systems.

5.1 Technological Dependence

The progressive integration of intelligent systems into preventive management brings with it the risk of creating a structural dependence on technology—one that should not be underestimated (Rodríguez Elizalde, 2025). This phenomenon goes beyond technical concerns and affects operational autonomy, human decision-making, and the ability to respond to unforeseen situations not contemplated by algorithms.

One of the most evident risks is overreliance on artificial intelligence systems, which can undermine professional judgment. When algorithms demonstrate high accuracy in repetitive tasks, there may be a temptation to delegate decision-making without critical review. This uncritical automation can lead to the gradual deactivation of human oversight and the erosion of technical skills among staff

(Torres Cadillo, 2023), especially if AI begins to be perceived as an autonomous decision-maker rather than a support tool.

The issue worsens if the system contains configuration errors, training data biases, or misinterprets the operational context. In such cases, if personnel are not equipped to question, validate, or intervene, the consequences can be serious (Goñi Sein, 2024). Therefore, it is essential to establish clear protocols for oversight, periodic validation, and the maintenance of meaningful human control—especially for critical decisions affecting occupational safety and health.

Another point of vulnerability is the dependence on external technological infrastructure, such as cloud services, constant connectivity, or remote maintenance. A service outage, cyberattack, or network failure could render the entire prevention system inoperative at crucial moments. To address this, organizations must implement contingency protocols, offline capabilities, and internal training to ensure prevention continuity during technological failures.

From a strategic standpoint, technological dependence also challenges the preservation of operational sovereignty. Delegating everything to external vendors without developing internal competencies may lead to a loss of know-how, lack of transparency in algorithms, and difficulty auditing the system's actual functioning.

In short, technological dependence is not an inevitable side effect—it is a manageable risk (Rodríguez Elizalde, 2025). The goal is not to slow innovation, but to ensure that its integration does not undermine the technical autonomy of the prevention system or reduce safety management to an opaque, automated process detached from professional judgment.

5.2 Privacy and Data Protection

One of the most sensitive areas in the application of artificial intelligence to occupational risk prevention involves the collection, processing, and use of personal data—particularly when such data can infer workers' physical, emotional, or behavioral states (Chicaiza Ortiz, 2024). As systems become more sophisticated, their technical potential increases—along with the ethical and legal challenges regarding privacy, dignity, and data protection.

Many AI systems applied to occupational risk prevention require highly personal information to function effectively. Examples already discussed in previous chapters (Rodríguez-Martínez et al., 2023) include:

- Biometric data, such as heart rate, fatigue, posture, or emotional state, obtained through wearable sensors or cameras (as described in Sections 3.2 Real-Time Monitoring and 3.6 Detection of Fatigue and Emotional States).
- Images and audio, captured by computer vision or intelligent surveillance systems (also in Sections 3.2 and 3.4 Computer Vision for Hazard Detection).
- Performance indicators, such as reaction times or equipment usage patterns, used to build behavioral profiles (Sections 3.4 and 3.6).

This type of data is protected under the General Data Protection Regulation (GDPR) in the EU, which classifies biometric and health-related data as especially sensitive. Their processing requires strict compliance with enhanced requirements regarding consent, transparency, proportionality, and data minimization (Torres Cadillo, 2023).

One of the main challenges is ensuring that worker consent is freely given, informed, specific, and revocable, which is not always straightforward in the context of a hierarchical employment relationship. Additionally, organizations must demonstrate that the use of such data is proportionate to the risk, necessary for the intended preventive purpose, and not replaceable by less invasive means.

Careful attention must also be paid to post-collection data use. Even if data is collected for a legitimate safety purpose—such as preventing fatigue-related accidents—there is a risk it could be repurposed to evaluate performance, influence promotions, or impose sanctions. Such use would violate fundamental principles of fairness and respect for the individual (Schor-Landman, 2022).

Without clear regulation and effective algorithmic governance, the door is left open to potential misuse—whether by the employer or by third parties. For this reason, the principle of privacy by design and by default must be applied from the outset, incorporating measures such as anonymization, encryption, access controls, internal audits, and oversight by data protection officers.

Privacy must not be handled solely by the legal or IT departments—it must be an integral part of the organization's preventive strategy and culture. A truly ethical AI-based occupational safety system must not only be technically efficient, but also legally compliant, organizationally transparent, and socially acceptable (Chicaiza Ortiz, 2024).

This debate becomes even more important with the growing presence of technologies such as biometric monitoring, intelligent video surveillance, and automated behavioral analysis in the workplace. In this context, the protection of fundamental rights must not be seen as a barrier—but as a prerequisite for ensuring that safety and privacy are not competing interests, but complementary pillars of a responsible prevention system.

5.3 Lack of Transparency (the "Black Box" Problem)

One of the most pressing challenges in the application of artificial intelligence to occupational risk prevention is the inherent opacity of many advanced models—particularly those based on deep learning. This phenomenon, known as the "black box" problem, refers to the difficulty of understanding how a system has arrived at a specific prediction, recommendation, or alert, even when its overall performance is objectively high.

This lack of explainability is not merely a technical issue: it compromises trust, effectiveness, and the governance of the preventive system. In contexts where decisions directly impact people's health and safety, it is not enough for artificial intelligence to be "accurate"—it must also be able to justify its outputs in a comprehensible, reasoned, and auditable manner (Antezana Soria Galvarro, 2023; Blasco Jover, 2024).

Below are five key risks arising from this lack of transparency:

1. Distrust in recommendations: When a system issues a high-risk alert, suggests task reassignment, or recommends a procedural change without explaining why, users tend to distrust—or outright ignore—the advice. In cognitively demanding environments where workers need to quickly distinguish between relevant warnings and non-essential signals, lack of clarity undermines the system's authority and contributes to "algorithmic fatigue," a phenomenon akin to alarm fatigue in clinical settings.
2. Difficulties with traceability and auditing: In the event of an incident or major failure, it is essential to reconstruct what happened, assign accountability, and verify whether protocols were followed. If the system acts as a black box, it is impossible to audit its internal logic—undermining continuous improvement and regulatory compliance. The lack of traceability also weakens the organization's legal position when justifying its preventive decisions.
3. Limiting meaningful human oversight: Meaningful human oversight—defined as the ability to review, understand, and correct decisions made by AI—is a basic ethical principle. Yet in opaque models, occupational health and safety professionals lose their capacity for critical intervention,

becoming passive recipients of decisions they cannot question or correct. This erodes professional judgment and context-based decision-making.

4. Educational and training challenges: AI systems that issue alerts without understandable justification limit the opportunity for experiential learning. Workers are unable to grasp the reason for the warning or which behavior triggered the risk, making sustained behavioral change more difficult. Opacity also fosters perceptions of arbitrary decision-making, further fueling distrust or resistance to change within the organization.

5. Risk of hidden errors and unjust decisions: If we don't understand how a model functions, we cannot anticipate or detect its failures. Opacity may conceal biases, errors, or spurious correlations that remain invisible until they cause a serious incident. If the AI is also used to segment profiles, evaluate performance, or recommend sanctions, lack of explainability can result in unfair or discriminatory decisions with no recourse for appeal.

These risks are not merely theoretical. In real industrial contexts, there have already been documented cases in which poorly trained algorithms produced false alarms (Zaidi et al., 2024) or ignored real risks by misinterpreting data (Greenemeier, 2018), even contributing to serious incidents. For this reason, there is growing consensus around the need to incorporate mechanisms of explainability and interpretability in all AI systems that affect human health or safety.

This means using models that are intrinsically more transparent—such as decision trees or rule-based systems—or applying explanatory tools (e.g., LIME, SHAP, or saliency maps) to generate representations that are understandable to human users when more complex models are used.

In short, it is not enough for artificial intelligence to "work"—it must be able to explain how and why it works. Algorithmic transparency is not a technical bonus; it is a fundamental requirement for the legal legitimacy, organizational acceptability, and operational effectiveness of the preventive system.

The black box problem should not be viewed as a technical oddity, but as a structural limitation that requires extremely cautious integration in high-stakes human environments. In a truly ethical and collaborative preventive system, explainability must be as much a priority as accuracy. It's not just about predicting risk—but managing it in a way that is understandable, responsible, and shared.

5.4 Bias Propagation

One of the most subtle yet potentially harmful risks in the application of artificial intelligence to occupational risk prevention is its tendency to reproduce and even amplify existing biases present in training data or model design. This phenomenon, known as bias propagation, can lead to unintended discrimination, systematic omissions, or decisions that—rather than improving equity—reinforce structural inequalities already embedded in the system (Chicaiza Ortiz, 2024; Schor-Landman, 2022).

At its core, artificial intelligence learns from the data it is given. If that data is tainted by historical, cultural, or organizational biases, the system internalizes and operates on them—even when the model performs "correctly" from a computational perspective.

In the field of occupational risk prevention, several types of bias can arise, affecting the system's effectiveness, fairness, and legitimacy:

- Representation bias: Occurs when the training data fails to adequately reflect the diversity of the workforce. For example, a computer vision system trained only on images of male workers in daytime industrial environments may struggle to interpret situations involving women, racialized individuals, or night shifts—resulting in omissions or false negatives in real-world scenarios.
- Underreporting bias: If certain groups, shifts, or departments have historically reported fewer incidents—due to fear, pressure, or organizational culture—AI may learn that those contexts pose less risk, even if real dangers are simply being concealed.
- Algorithmic bias: Arises when the model gives undue weight to variables such as age, gender, or job role, generating spurious correlations between personal characteristics and risk levels. This can result in unfair evaluations or disproportionate preventive measures.
- Structural bias: Appears when certain groups, such as subcontractors or temporary workers, are excluded from data analysis and system coverage. This leads to incomplete protection and unmanaged residual risk.

The consequences of such biases are significant and concerning (Alawamleh et al., 2024):

- Invisibility of vulnerable groups, whose risks go undetected or unaddressed.

- Inefficient or unfair resource allocation, favoring historically visible areas or profiles over more exposed but underrepresented ones.
- Perceptions of arbitrariness or discrimination, which can erode organizational trust and foster internal resistance.
- Reinforcement of dysfunctional safety cultures, where lack of reporting is interpreted as "good performance."

Fortunately, this risk is manageable through deliberate strategies incorporated into system design:

- Data audits, ensuring that training datasets reflect workplace diversity (gender, age, experience, environments, shifts, etc.).
- Algorithmic fairness testing, to detect disproportionate impacts and apply corrective techniques (reweighting, normalization, filtering).
- Explainable and traceable models, enabling the identification and critical review of biased decisions.
- Participatory design, involving safety technicians, union representatives, diverse profiles, and ethics experts to review assumptions and criteria.
- Algorithmic impact assessments prior to deployment—especially in systems that could affect performance evaluations, task assignments, or disciplinary measures.

The following table summarizes common types of bias in AI applied to occupational risk prevention, along with examples and associated operational or ethical risks:

Table 11. Examples of Bias in AI Applied to Occupational Risk Prevention (table by the autor).

Type of Bias	Example in OSH	Associated Risk
Representation bias	System trained on images of male workers during daytime shifts.	Omission of risks involving non-standard profiles or unrepresented conditions.
Underreporting bias	Low reporting in high-pressure shifts or departments with a culture of silence.	False perception of safety; invisibility of critical zones.
Algorithmic bias	Erroneous link between young age and high risk without context.	Disproportionate or discriminatory preventive measures.
Spurious correlation bias	Interpreting high reporting rates as danger instead of good safety culture.	Penalizing good practices; misallocation of resources.
Structural bias	Exclusion of subcontracted staff from risk analysis.	Incomplete system coverage; increase in unmanaged residual risk.

Artificial intelligence is not neutral (Chicaiza Ortiz, 2024). It reflects—and often amplifies—the values, omissions, and assumptions encoded in its data and design. In a safety culture where equity, inclusion, and organizational justice are essential, allowing AI to operate with unchecked biases jeopardizes both the system's ethical legitimacy and technical effectiveness.

Therefore, integrating AI into occupational risk prevention must be accompanied by a critical and systemic lens, ensuring that technological progress does not result in new forms of inequality—but rather enables a fairer, more transparent, and diversity-sensitive prevention system.

5.5 Implementation Cost

The integration of artificial intelligence (AI) into occupational risk prevention systems is not merely a technological decision—it is, above all, a strategic investment. However, taking this step poses a significant economic and operational barrier, especially for organizations with traditional structures, low levels of digitalization, or limited resources.

The costs associated with implementing AI in preventive environments go far beyond software alone and must be assessed holistically. Key elements to consider include:

- Specialized hardware: smart cameras, IoT sensors, local or cloud servers, wearables for physiological monitoring, etc.
- Software and licenses: commercial solutions for computer vision, predictive analytics, virtual assistants, etc., which require licensing, maintenance, and updates.
- Technical consulting and integration: adaptation to the company's context, model training, functional customization, and integration with existing systems.
- Staff training: equipping occupational health and safety teams, operators, and supervisors to interpret alerts, use interfaces, and understand system logic.
- Organizational adaptation: redesign of workflows, action protocols, and reporting systems to align with the new digital tools.
- Maintenance and scalability: regular algorithm reviews, performance monitoring, and updates to keep pace with technical or regulatory changes.

In large or digitally mature companies, these costs can be absorbed as part of broader digital transformation strategies. However, for many SMEs—particularly in sectors like construction or auxiliary industries—the initial cost may appear disproportionate to the expected short-term benefits, thereby widening the digital divide in workplace safety between large and small organizations.

Nonetheless, the key is to view this investment from a medium- and long-term return perspective. When implemented correctly, AI solutions can:

- Reduce accident rates and associated costs from sick leave, penalties, or legal claims.
- Minimize operational downtime through predictive maintenance.
- Optimize the use of safety resources by focusing efforts where they are most needed.
- Improve traceability, auditability, and regulatory compliance.

Furthermore, public support programs (such as European digitalization funds or national technology-promotion initiatives) can help finance these solutions for smaller companies.

The table below summarizes the main implementation costs of AI in workplace safety alongside the expected operational benefits:

Table 12. Implementation Costs vs. Operational Benefits of AI in OSH (table by the author).

Assessed Aspect	Estimated Cost (initial or ongoing)	Expected Operational Benefit
Specialized hardware	High (depending on system size and complexity)	Capturing critical data and automating monitoring
Software and licenses	Moderate to high (depending on provider and features)	Intelligent processing, automation of complex tasks
Consulting and integration	Moderate (initial phase, integrating AI with existing systems)	Customization, alignment with risks and organizational culture
Staff training	Moderate (one-time or recurring, depending on turnover)	Improved adoption, usability, and system effectiveness
Maintenance and updates	Ongoing (essential for sustained performance and relevance)	Robust system adapted to technical and regulatory changes
Accident reduction	Investment with progressive return in the medium term	Fewer incidents, savings on compensation, enhanced corporate image
Reduced downtime	Immediate benefit when predictive monitoring is implemented	Increased operational continuity, fewer losses from interruptions
Optimization of OSH resources	Direct operational return when resources are aligned to high-risk areas	Efficient use of personnel, inspections, and preventive actions
Automated regulatory compliance	Indirect savings in the medium term (fewer penalties or inspections)	Improved legal compliance, reduced exposure to regulatory risks
Improved traceability and reporting	Continuous cross-cutting benefit	Data-driven decision-making, smoother internal and external audits

The cost of implementing AI in occupational safety should not be viewed as a prohibitive barrier, but rather as a lever for modernizing the prevention system. The challenge for each organization is to develop a realistic, phased roadmap tailored to its context, prioritizing modular, interoperable solutions focused on measurable outcomes.

Only in this way can we prevent technological innovation from reinforcing existing inequalities and ensure that AI becomes an accessible and beneficial tool for the entire productive fabric, not just for large corporations.

5.6 Rejection and Distrust Among Personnel

The implementation of artificial intelligence in the workplace—particularly in the field of occupational risk prevention—requires more than technical and organizational transformation. It also involves a reconfiguration of the relationships between technology, people, and trust. One of the most relevant and often underestimated barriers is the rejection or distrust that these tools can generate among workers, especially when their introduction is not accompanied by appropriate strategies for communication, training, and participation.

The resistance to AI in the workplace stems from multiple dimensions, both rational and emotional (see also Section 5.2 on privacy and data protection):

- Feeling of constant surveillance: The use of smart cameras, biometric sensors, or behavioral analysis systems may be perceived as a form of control rather than a preventive measure, leading to discomfort, stress, and a sense of intrusion into privacy.
- Fear of automated decision-making: When algorithms intervene in task assignments, fatigue-based break validation, or behavioral assessments, workers may feel a loss of autonomy or that professional judgment is being replaced.
- Job security concerns: Even though AI in occupational safety is conceived as a supportive tool, in highly automated contexts it may trigger fear of job replacement—especially if the complementary role of technology is not clearly communicated.
- Lack of algorithmic transparency: If system decisions are not explainable or understandable, workers tend to distrust its alerts and recommendations, undermining the system's credibility and acceptance.
- Lack of active participation: When technology is introduced without consultation or real involvement of those who will use it, it may be viewed as a top-down imposition disconnected from the actual work environment.

This type of rejection is not anecdotal nor trivial. It can directly impact the effectiveness of the preventive system in the following ways:

- Reduced collaboration during implementation and daily use.
- Intentional tampering or sabotage of sensors and cameras.
- Underreporting of incidents due to fear of consequences.
- Passive resistance that slows down or blocks deployment.

Ultimately, AI that is not accepted is AI that is ineffective. Organizational trust is just as crucial as technical accuracy in ensuring the sustainability of technological change.

Overcoming this barrier requires a holistic approach that combines technical, ethical, communicative, and participatory dimensions. Key strategies include (Niebles, 2020; Rodríguez Elizalde, 2025):

- Transparency from the outset: Clearly explain how the system works, what data it collects, its purposes, and who has access to and control over the data.
- Legal safeguards and informed consent: Ensure that privacy rights, data protection, and non-discrimination are fully respected.
- Tailored training: Provide all staff—whether low-skilled workers or technical professionals—with the necessary knowledge to understand the real safety and health benefits of AI.
- Meaningful human oversight: Ensure that critical decisions are subject to professional review, and that workers can challenge or add context to algorithmic recommendations.
- Participation from the design phase: Involve staff early (e.g., pilots, validations, field testing) to facilitate ownership, gather feedback, and tailor the system to actual work dynamics.

Table 13 summarizes the main common sources of rejection and the corresponding mitigation strategies:

Table 13. Acceptance Barriers and Mitigation Strategies (table by the author).

Acceptance Barrier	Impact on the Work Environment	Mitigation Strategy
Feeling of constant surveillance	Stress, punitive control perception, deteriorated trust climate	Define clear limits; ensure use is strictly preventive
Fear of automation	Rejection of recommendations, delegitimization of the system	Reinforce human oversight; communicate technology's supportive role
Fear of job loss	Resistance to use, perception of professional obsolescence	Emphasize that AI complements rather than replaces people; highlight expert judgment
Lack of technological transparency	Distrust in automated alerts or decisions	Present decision criteria clearly; enable traceability
Absence of active participation	Low engagement, lack of ownership, informal resistance or sabotage	Involve workers from early stages; tailor the system to their real work experience

In summary, the success of artificial intelligence in occupational risk prevention does not depend solely on technical accuracy, but on its ability to earn the trust of the people who must work alongside it (Otero Mateo et al., 2023; Ruiz & Velásquez, 2023). A workplace where technology is perceived as a threat, an imposition, or a surveillance mechanism is doomed to fail. But when it is presented as a tool for collective safety, introduced with transparency, dialogue, and mutual respect, AI not only enhances safety—it also improves workplace climate, organizational trust, and openness to change.

5.7 Risk of Technological Obsolescence

One of the most significant (and yet often least visible) risks associated with the implementation of artificial intelligence in occupational risk prevention is technological obsolescence (Mellal, 2020). Unlike other more stable technical solutions, AI systems are especially vulnerable to the passage of time due to the accelerated pace of evolution in algorithms, technological platforms, regulations, and operational environments.

An AI solution implemented today may lose effectiveness—or even become unusable—within just a few years (or even months), unless supported by a clear strategy for continuous updating, maintenance, and evolution. Some of the main factors influencing this obsolescence include (Leguizamón Páez & Becerra Correa, 2024):

- Advances in models and architectures: What is currently standard (e g., convolutional networks or transformers) may soon be surpassed by more precise, explainable, or energy-efficient models.
- Dependence on external tech ecosystems: Frameworks, libraries, or cloud platforms may stop receiving updates, become incompatible, or be discontinued.
- Regulatory changes: New regulations like the European Union Artificial Intelligence Act (AI Act), or evolving standards on data protection, cybersecurity, and occupational health, demand constant adaptation to maintain compliance.
- Variations in operational conditions: Changes in processes, tools, shifts, or workforce profiles may render outdated models inaccurate unless retrained and adjusted.
- Cybersecurity risks: Systems without regular maintenance or updates can become vulnerable to unauthorized access or external manipulation.

Technological obsolescence does not only impact accuracy—it can result in operational, economic, legal, and reputational risks (Mendizá & Jiménez, 2021):

- Decreased reliability of alerts or recommendations.
- Additional costs from system migrations or replacements.
- Regulatory non-compliance leading to penalties.
- Security breaches due to unpatched vulnerabilities.
- Poor scalability or difficulty integrating new tools.

To mitigate this risk, organizations must adopt a long-term vision, understanding that AI is not a plug-and-play solution, but a living system that requires ongoing care. Recommended best practices include (Goñi Sein, 2024; Ruiz & Velásquez, 2023):

- Planning the technological life cycle in advance, including periodic reviews and phased upgrades.

- Choosing modular, open, and interoperable technologies that allow for incremental improvements without complete system overhauls.
- Including evolutionary support clauses in contracts with tech providers to ensure continuous maintenance, regulatory adaptation, and functional updates.
- Investing in the continuous training of internal teams to enable proper supervision, evaluation, and contextual system adaptation.
- Performing regular technological audits, to assess model performance, data quality, and legal compliance every 12 to 24 months.

Table 14. Risks of Technological Obsolescence and Mitigation Strategies (table by the author).

Obsolescence Factor	Associated Risk	Recommended Mitigation Practice
Obsolete algorithmic models	Loss of system accuracy and reliability	Periodic retraining; architecture review every 1–2 years
Outdated or discontinued software	Incompatibility; feature loss	Contracts with active support; use of tools with strong developer communities
Legal or regulatory changes	Legal non-compliance; risk of penalties	Regulatory monitoring; alignment with frameworks like the AI Act
Evolution of the work environment	Inaccurate outputs; reduced operational value	Functional audits; user feedback collection
Cybersecurity vulnerabilities	Data leaks or system manipulation	Patch management; periodic infrastructure reviews

In conclusion, technological innovation in occupational risk prevention should not be understood as a one-time action, but rather as a continuous process of adaptation, evaluation, and improvement. Obsolescence is an inherent risk in any artificial intelligence system, but it is not inevitable: with strategic planning, organizational commitment, and a preventive mindset, it is possible to ensure that these solutions remain up-to-date, secure, and aligned with the goals of effective, sustainable, and evidence-based prevention.

5.8 Ethics of Artificial Intelligence in Workplace Environments

Beyond the technical, functional, or legal risks associated with the use of artificial intelligence in occupational risk prevention, it is essential to consider its ethical dimension. The incorporation of these technologies in human environments must guarantee principles such as dignity, autonomy, justice, equity, and transparency, especially when automated decisions directly affect workers' safety, health, or rights.

5.8.1 Respect for Workers' Dignity and Autonomy

Risk prevention should not be limited to ensuring safe environments, but also to preserving the dignity and personal integrity of those performing their work. Misguided use of AI can undermine this principle when:

- Professional judgement is replaced without effective human oversight.
- Surveillance mechanisms are perceived as invasive.
- Automated decisions are applied without the possibility of review or appeal.

Every AI-based prevention system must guarantee the principle of meaningful human oversight, i.e., any critical decision must be understandable, controllable, and validated by competent individuals.

5.8.2 Equity and the Fight Against Algorithmic Discrimination

The reproduction of biases in AI systems is one of the most pressing ethical concerns. Using historically biased data, lacking representativeness, or designing inappropriate algorithms can lead to:

- Unfair personalized risk assessments.
- Errors in judgement based on accent, age, gender, or physical appearance.
- Technological exclusion of profiles that do not match the learned model.

Algorithmic fairness requires auditing training data, validating system decisions, and implementing continuous review mechanisms to detect and correct unintended discrimination.

5.8.3 Transparency, Explainability, and Traceability

An ethical system must explain clearly how it works and why it makes certain decisions. This is especially important in prevention, where trust in alerts is key:

- Technicians must understand the criteria that trigger a risk alert.
- Workers must know how they are being evaluated and what the consequences are.
- Every automated decision must leave a traceable record for review in case of incident or audit.

Without transparency or traceability, responsibility becomes diluted, and control over the system is compromised.

5.8.4 Consent and Right to Information

When AI systems process personal or sensitive data (e.g., biometrics, fatigue, or behavioral patterns), they must comply not only with data protection laws, but also with the ethical obligation to ensure informed, voluntary, and revocable consent.

This means workers must:

- Know what data is being collected, why, and what the consequences of its use may be.
- Be able to refuse or limit data use without retaliation.
- Not be subject to exclusively automated decisions of significant impact, as established by Article 22 of the GDPR.

5.8.5 Organizational Sustainability and Digital Prevention Culture

An ethical AI implementation requires a mature preventive culture, where technology complements human responsibility, rather than replacing or weakening it. This requires:

- Active worker participation in the design, validation, and implementation phases.
- Training for prevention professionals not only in technical or regulatory aspects, but also in ethical dimensions.

- Evaluation of both the system's effectiveness and its impact on motivation, trust, and organizational climate.

In short, applying AI to occupational risk prevention is not just about fulfilling technical or legal requirements, but about embracing a solid ethical commitment, placing the worker at the center of the system's design, deployment, and supervision.

Table 15 outlines the key ethical principles that should guide the responsible use of artificial intelligence in occupational safety environments.

Table 15. Ethical Principles Applicable to AI in Workplace Prevention Environments (table by the author)

Ethical Principle	Description	Practical Application in OSH
Human Oversight	AI must not operate autonomously without meaningful human validation.	Manual validation of risk alerts; oversight of automated decisions.
Transparency & Explainability	System decisions must be understandable and auditable.	Show algorithm criteria to OSH staff; justify automatic decisions.
Fairness & Non-Discrimination	AI must not perpetuate biases or exclude vulnerable groups.	Review training data, audit for biases, include gender and age perspectives.
Consent & Right to Information	People must know how their data is used and be able to decide on it.	Clearly inform about camera/sensor use or behavioral analysis.
Traceability & Logging	Verifiable records of system functioning and decisions must exist.	Log alerts, automated decisions, and human corrections in OSH system.
Participation & Social Dialogue	Implementation must include worker and union representative input.	Involve OSH committees, OSH technicians, and prevention delegates from early stages.
Technological Proportionality	Avoid using AI when simpler, more effective means are available.	Prevent excessive systems for minor risks or overengineered solutions.
Digital Preventive Culture	AI should reinforce prevention, not replace or weaken professional judgement.	Complement human expertise; avoid replacing OSH professionals with systems.

In conclusion, ethics is not an optional add-on in the application of AI to occupational safety, but a structural foundation. Only through a human-centered approach, based on transparency, fairness, and participation, will it be possible to build technological systems that are not only effective, but also legitimate, sustainable, and socially accepted. Trust, like safety, cannot be automated — it must be cultivated.

6 Sectoral Approach: Artificial Intelligence in Construction Risk Prevention

6.1 Specific Features of the Construction Sector

The construction industry is one of the most complex and demanding sectors in terms of occupational risk prevention. Its inherently dynamic, fragmented, and ever-changing operational nature makes construction sites highly uncertain environments where risks are not only numerous but also fluctuate continuously and are difficult to standardize.

Unlike more stable and controlled industrial contexts, construction activity presents a set of specific features that decisively influence the planning, implementation, and monitoring of preventive measures. The most relevant include:

- Non-standardized and constantly evolving work environment. Construction sites are temporary, open systems in continuous transformation. As work progresses, fronts, access points, structures, and surrounding conditions change, requiring adaptive, real-time safety surveillance.
- High concentration of hazardous activities and concurrent operations. Working at height, excavation, use of heavy machinery, structural assembly, and load handling are intrinsically hazardous tasks. Their simultaneous execution in different zones and elevations increases the likelihood of interferences and cross-accidents, such as falling objects, entrapment, or electrical contacts.
- High degree of subcontracting and business fragmentation. Construction sites often involve multiple contractors and subcontractors, each with its own safety culture, training level, and procedures. This organizational diversity complicates coordination and requires interoperable management and control systems. AI can contribute by enabling traceability, standardization, and automatic detection of deviations.
- Diverse and high-turnover workforce. Construction crews are often highly heterogeneous in terms of experience, language, training, and contractual status. This diversity, combined with frequent turnover, makes it challenging to monitor training, evaluate cumulative fatigue, or verify technical competencies. AI applied to personalized risk management

enables individualized and dynamic traceability, even in highly mobile environments.

- Exposure to adverse environmental conditions. Many construction tasks take place outdoors or in partially sheltered spaces, exposing workers and structures to wind, rain, extreme heat, or low visibility. These conditions affect not only physical safety but also cognitive performance and reaction capacity.
- Time and productivity pressures. Tight deadlines and budget constraints can generate tensions between production goals and safety. In such contexts, it is not uncommon for safety controls to be relaxed, shortcuts taken, or performance prioritized over safe conditions. Automating preventive controls through AI enables increased surveillance without adding operational workload.
- Technological and connectivity limitations. Many sites, particularly in early phases or remote areas, lack the infrastructure needed for digital risk management. This necessitates the design of robust, portable, scalable technological solutions capable of operating under low connectivity and limited power supply.

Taken together, these characteristics make construction a highly complex safety environment requiring flexible, predictive, and personalized approaches. Artificial intelligence thus emerges not merely as a complementary tool but as a structural component for transforming the prevention model, enabling:

- Anticipation of risks before they materialize.
- Automation of real-time controls without continuous supervision.
- Coordination across fragmented organizational environments via shared platforms.
- Personalization of safety measures according to each worker's profile, task, and condition.
- Reinforcement of traceability, adaptive training, and data-driven decision-making.

To be truly effective, construction risk prevention must move beyond reactive paradigms and embrace a logic of intelligent anticipation, where AI acts as a continuous, operational, and strategic support across all levels of the safety management system.

6.2 Key Accident Statistics

A quantitative analysis of occupational accident rates in the construction sector is a crucial tool for supporting preventive decisions, prioritizing resources, and identifying patterns of structural risk. Over recent decades, data collected by organizations such as the International Labour Organization (ILO), the European Agency for Safety and Health at Work (EU-OSHA), and, at the national level, the Spanish National Institute for Occupational Safety and Health (INSST) consistently indicate that construction remains one of the sectors with the highest incidence of serious and fatal workplace accidents (European Agency for Safety and Health at Work, 2022; Instituto Nacional de Seguridad y Salud en el Trabajo, 2024; International Labour Organization, 2023).

Far from being merely a statistical reality, these figures reflect the persistence of structural risk conditions that are difficult to correct through conventional preventive approaches. It is precisely in these limitations that the transformative potential of artificial intelligence lies: by combining sensorization, predictive analysis, and digital traceability, it enables data to be transformed into operational knowledge and monitoring into a continuous, adaptive, and proactive function.

6.2.1 Disproportionate and Persistent Incidence

At the European level, the construction sector represents around 5–8% of total employment but accounts for over 20% of reported work-related fatalities. This disproportion, consistent over time, highlights the presence of systemic risk factors that remain unresolved.

In Spain, the annual report by the INSST (Instituto Nacional de Seguridad y Salud en el Trabajo, 2024) notes that:

- The incidence rate was 80% higher than the average across all other sectors.
- The severity index was 40% above the national average, highlighting the greater seriousness of injuries.
- The mortality rate exceeded 7 deaths per 100,000 workers—nearly triple the cross-sector average.

These data are repeated across various autonomous regions and types of construction work, confirming that this is not a circumstantial anomaly but rather a widespread and persistent pattern deeply rooted in the sector's inherent conditions.

6.2.2 Most Common Causes of Workplace Accidents

A detailed analysis of recorded accidents helps identify the most recurrent types and their predominant causal factors:

- Falls from height, the leading cause of fatalities, often resulting from inadequate or absent use of harnesses, lifelines, or guardrails.
- Struck by moving objects or falling materials, worsened by the simultaneous execution of tasks at different heights or by inefficient logistics.
- Being caught or crushed by machinery or structures, such as backhoes, heavy vehicles, or formwork and shoring systems.
- Electrical contacts, both due to poor temporary installations and accidental drilling in unmarked zones.
- Musculoskeletal injuries, caused by overexertion, awkward postures, or repetitive tasks.
- Human errors induced by fatigue or stress, which, although rarely listed as the direct cause, are present as contributing factors in a significant proportion of incidents.

6.2.3 Operational and Temporal Patterns

Temporal analysis of the data also reveals several recurring patterns:

- Higher accident rates in the final hours of a shift, due to accumulated fatigue and decreased alertness.
- Increased incidents on Monday mornings, linked to physical and cognitive adjustment after rest days.
- Elevated accident rates during critical project phases, such as closures or handovers, where multiple tasks converge and deadline pressure intensifies.
- Greater vulnerability among newly hired or inexperienced workers.

These patterns highlight the need for preventive strategies that not only react to incidents but also anticipate periods and profiles of heightened risk exposure.

6.2.4 Shortcomings Revealed by Accident Data

Longitudinal analysis of incident records reveals several structural weaknesses in current prevention systems:

- Lack of continuous supervision and real-time response capability.
- Discrepancy between standardized procedures and actual working conditions.
- Poor integration of data between contractors and subcontractors, hindering a comprehensive view of risk.
- Limited capacity to predict human behavior under pressure, fatigue, or multitasking scenarios.

These shortcomings open the door for artificial intelligence technologies to act as drivers of improvement. Systems that can learn from real data, recognize patterns of deviation, and generate real-time preventive alerts can help convert statistics into operational safety.

6.2.5 AI Applications for the Most Common Types of Accidents

Based on the most frequent types of incidents, various artificial intelligence-based solutions have been identified with high preventive potential. Table 16 summarizes the main accidents, their causes, and corresponding technological applications:

Table 16. Common construction accidents and AI-based applications (table by the author)

Type of Accident	Main Causes	AI-Based Solutions
Falls from height	Improper or absent use of fall protection systems	Computer vision to verify PPE use; detection of presence in elevated unprotected areas
Struck by moving objects	Poor logistics; unsecured loads; poorly coordinated tasks	AI-powered cameras and sensors to predict object trajectories and fall risk
Caught/crushed by machinery	Maneuvers in blind spots; unmarked machinery	Proximity algorithms; route prediction; immediate alerts for unexpected movements
Electrical contacts	Lack of prior planning; unawareness of underground systems	Smart plan analysis; AI-based geolocation to identify electrical hazard zones
Muscular overexertion	Excessive physical load; awkward or repetitive postures	Biometric sensors; posture and fatigue analysis using machine learning
Human errors due to fatigue	Long shifts; extreme heat; organizational pressure	Facial/voice recognition AI; real-time task modulation and break recommendations

This approach, grounded in concrete data and applied technology, is not intended to replace professional judgment, but rather to enhance and expand it through tools that increase the detection, anticipation, and adaptability capacities of the preventive system.

When statistics are combined with intelligent analysis and context-aware action models, they become a lever for preventive transformation. The following section will present functional examples of these applications, aimed at specifically addressing the structural risks inherent to the construction sector.

6.3 AI-Based Management of Specific Occupational Risks

Construction is an environment where risks are not only numerous, but also highly specialized. Each type of hazard stems from specific technical, organizational, and human causes that require tailored—and often individualized—preventive measures. In this context, artificial intelligence (AI) emerges as a versatile and scalable tool capable of anticipating, detecting, and mitigating risks in real time by integrating multiple sources of information and contextual analysis.

Below is a detailed overview of the main specific construction-site risks and their potential mitigation through intelligent technologies (Instituto Nacional de Seguridad y Salud en el Trabajo, 2024; Luna López et al., 2021; Monereo Pérez & López Insua, 2016; Muñoz Perez et al., 2022; Pampliega, 2019):

- Falls from height: The leading cause of fatal accidents on construction sites, often due to the absence or failure of collective protections (guardrails, safety nets), improper harness use, missing lifelines, or unmarked access points. AI can address these risks through:
    - Computer vision to verify the correct use of fall protection equipment such as harnesses.
    - Automated video analysis in critical zones (walkways, platforms, edges).
    - Predictive models that identify high-risk areas based on site geometry and progression.

  Practical example: A system of AI-powered cameras detects a worker on an elevated platform without a harness or with an improperly attached anchor point and issues an alert (Fig. 17 (a)).

- Crushing and heavy machinery-related hazards: This includes collisions with moving equipment, being caught between structural elements, or errors in machinery operation. AI solutions include:
    - Person-machine proximity detection via computer vision or LIDAR sensors.
    - Trajectory prediction and dynamic exclusion zone setting.
    - Embedded AI systems that adjust speed or trigger automatic braking.

  Practical example: Smart machinery equipped with AI vision detects workers in blind spots and halts operation if a minimum safety distance is breached (Fig. 17 (b)).

- Fatigue, distraction, and human error: Factors such as dehydration, heat stress, or production pressure increase the likelihood of operational mistakes. AI can help by:

- Wearables (goggles, helmets, wristbands) that monitor biometric signs: heart rate, microexpressions, blinking, or pupil dilation.
- Algorithms that detect attention lapses through behavioral patterns.
- Virtual assistants that assess cognitive status through interaction.

Practical example: Smart glasses tracking eye movements detect sustained visual fatigue and pause the operation to prevent further risk (Fig. 17 (c)).

- Unsafe environmental conditions (structural or weather-related): This includes unstable temporary structures, extreme weather, dust accumulation, or excessive vibrations. AI applications include:
  - IoT sensors to monitor tilt, vibration, or overload conditions.
  - Predictive models correlating weather data with construction phases.
  - Automated alerts or access control activation under critical conditions.

  Practical example: AI analyzes scaffold inclinometer data and triggers evacuation protocols when abnormal shifts are detected (Fig. 17 (d)).

- PPE (Personal Protective Equipment) management: Poor use, absence, or deterioration of PPE significantly compromises the effectiveness of the overall preventive system. AI enables:
  - Visual recognition to verify complete and correct PPE usage.
  - Smart access control allowing entry only when required PPE is detected.
  - Automated compliance logging.

  Practical example: An AI-enabled access gate scans the worker upon entry and denies access if any mandatory PPE element is missing (Fig. 18).

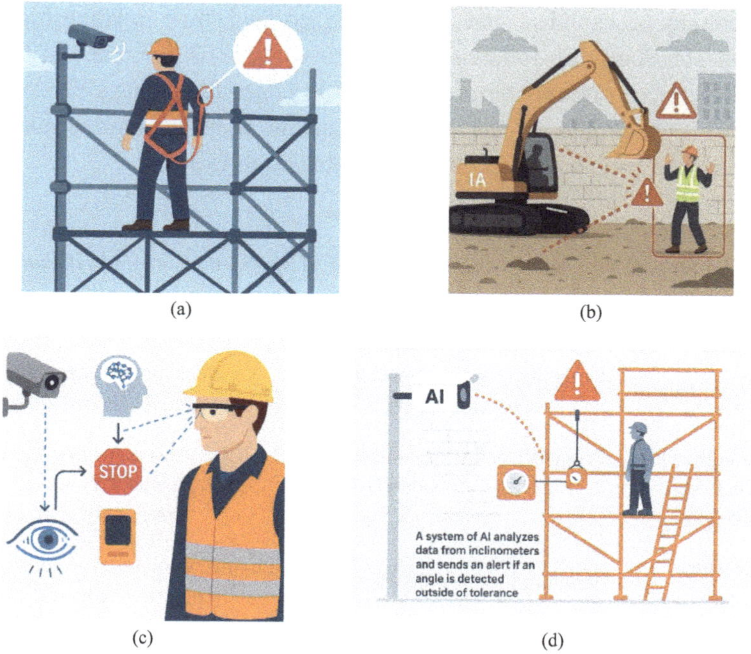

Fig. 17. Representations of AI Applications Addressing Specific Construction Sector Risks: (a) fall from height; (b) machinery entrapment; (c) visual fatigue; (d) structural instability (graphics by the author).

Fig. 18. PPE compliance verification through AI at restricted access point (graphics by the author).

To structurally consolidate the risks analyzed and their relationship with intelligent solutions, the following comparative table is presented. Table 17 summarizes the most common critical risks on construction sites, the artificial intelligence technologies applicable to each, and the added preventive value they offer in terms of anticipation, control, and reduction of exposure.

Table 17. Frequent Accidents in Construction and AI Applications (table by the author).

Specific Risk	Applied AI Technology	Preventive Value
Falls from height	Computer vision, geolocation, contextual alerts	Prevents unsafe access to unprotected elevated areas
Caught-in/between incidents	Proximity detection, trajectory prediction, embedded AI systems	Minimizes collisions and entrapments with heavy machinery
Structural instability	Structural IoT + predictive AI	Anticipates collapses or failures in temporary supports
Heat stress and fatigue	Biometric wearables, cognitive-behavioral analysis	Reduces human error due to fatigue or accumulated stress
PPE compliance	Automated visual recognition, smart access control	Ensures regulatory compliance and prevents unsafe working conditions

Taken as a whole, artificial intelligence enables the construction sector to address its specific risks through an adaptive, continuous, and personalized logic. Far from replacing professional judgment, these solutions reinforce it by expanding its capacity for monitoring, anticipation, and response. Their proper integration should always be framed within a broader preventive strategy, in which technology acts as an operational ally to occupational safety professionals (Baduge et al., 2022).

The following section will explore the advantages these systems offer in terms of control, coordination, and efficiency within complex environments such as construction sites.

6.4 Specific Advantages for the Construction Sector

The integration of artificial intelligence systems into occupational risk prevention does not merely represent an incremental improvement, but rather a qualitative leap forward for a sector as complex and demanding as construction. The inherent characteristics of construction sites—such as high operational variability, business fragmentation, and elevated exposure to risk (Fernández-Costales Muñiz, 2020)—make AI particularly valuable, offering specific, tangible, and strategic benefits that are difficult to achieve using traditional methods.

Below are the main differentiated advantages that artificial intelligence brings to the construction environment (Luna López et al., 2021; Muñoz Perez et al., 2022; Pampliega, 2019):

1. **Effective reduction of serious and fatal accidents**: Through its capabilities for continuous monitoring, automated detection, and predictive analysis, artificial intelligence can anticipate hazardous conditions before they materialize into incidents. This is especially effective for high-risk tasks such as working at height, operating heavy machinery, or intervening in unstable environments. Intelligent systems have proven useful in minimizing exposure times, preventing unauthorized access, and detecting operational deviations in advance.
2. **Enhanced operational control in complex environments:** In large-scale construction projects, where multiple activities, teams, and subcontractors coexist, manual safety oversight becomes unfeasible. AI-based systems enable widespread and ubiquitous supervision: strategically deployed cameras, sensors, and algorithms that detect incidents, verify safety conditions, and issue real-time alerts—even in the absence of onsite safety personnel.
3. **Facilitation of regulatory compliance:** Ensuring rigorous and documented compliance with legal occupational safety requirements is one of the main challenges in construction. Artificial intelligence supports this through:
    - Automated verification of proper use of personal protective equipment via computer vision.
    - Traceable digital recording of training sessions, inspections, site entries, and environmental conditions.
    - Automated generation of reports, checklists, and audit documentation. This not only reduces administrative burdens but also strengthens the company's legal position in the event of inspections or litigation.

4. **Coordination of concurrent activities**: In settings where dozens of companies operate simultaneously, artificial intelligence allows for centralized and standardized safety information management through shared platforms. These platforms enable:
    - Real-time visibility of each subcontractor's compliance with occupational safety requirements.
    - Detection of operational interference zones between teams.
    - Cross-alerting and automated coordination of collective protection measures.
      This fosters an integrated, collaborative, and data-driven safety management approach.
5. **Efficiency in preventive resource allocation:** AI enables intelligent allocation of technical and human resources based on the actual risk level of the environment. It allows:
    - Detection of zones with higher accident probability using predictive analytics.
    - Activation of virtual inspections focused on areas with anomalous behavior.
    - Reallocation of efforts toward more dangerous or less controlled project phases. This prevents resource dilution and maximizes the operational impact of preventive actions.
6. **Promotion of a Data-Driven Preventive Culture:** Artificial intelligence enables a shift from a reactive model—focused on post-incident correction—towards a proactive and educational approach. Thanks to real-time alerts, traceable interactions, and automated reports, prevention becomes a daily, visible, understandable, and measurable practice. This transformation reinforces the safety culture, improves system perception among workers, and encourages greater collective engagement.

In short, the advantages of artificial intelligence in construction are neither abstract nor generic: they directly respond to the structural challenges of the sector, delivering practical solutions where traditional methods fall short. Smart technologies do not replace traditional safety practices—they enhance and professionalize them, adding operational capacity, predictive insight, and real-time decision support (Luna López et al., 2021; Muñoz Perez et al., 2022; Toro

Lazo, 2024). A new model of prevention is indeed possible: more effective, more adaptive, and more aligned with the dynamic reality of construction sites.

6.5 Real Cases and Pilot Projects in Construction Companies and Sites

As developed in the previous sections, the integration of artificial intelligence systems in occupational risk prevention in the construction sector has moved beyond theoretical proposals to become a tangible, assessable, and, in many cases, replicable reality. Various construction companies, public entities, and technology startups have launched pilot projects and real-world applications that allow for on-site verification of the benefits, challenges, and necessary conditions for the effective implementation of AI in construction environments.

Below are some of the most representative national and international cases, selected for their demonstrative value and technical applicability:

- Skanska (Sweden): Automated PPE Compliance Monitoring via Computer Vision: The Swedish multinational deployed computer vision systems trained to recognize personal protective equipment (PPE) such as helmets, high-visibility vests, and fall protection harnesses on large construction sites (Kalogianni et al., 2021; Vuorinen et al., 2024).

  Key features of this case (A. Jones, 2024):

  - Use of fixed and mobile cameras with edge computing and AI models trained to recognize multiple clothing combinations and their placement.
  - Significant reduction in unauthorized access to restricted areas and sustained improvement in PPE compliance.
  - The project's standout value was the complete automation of compliance monitoring, even during night shifts, with no human supervision required.

- Smartvid.io (USA): Automated Image and Video Analysis on Construction Sites: The U.S.-based startup developed a platform capable of processing thousands of on-site images to automatically detect unsafe working conditions (Massiris et al., 2021).

  Key features (Park et al., 2020):

  - Used by major contractors such as Suffolk and Barton Malow.

    - Technology based on convolutional neural networks combined with continuous learning models, including human feedback loops.
    - In projects where the tool was actively used, incident frequency decreased by up to 20%, representing a substantial reduction in accident rates.

- Japan's Public Works: Autonomous Robots and AI-Driven Technical Monitoring: Japan's public infrastructure projects (e.g., tunnels, dams, and linear works) integrated AI-powered robotic systems for inspection and monitoring without direct human intervention (Pérez Herrero, 2023; Tirado Robles, 2022).

  Key features:

    - Use of integrated technologies for environmental sensing, thermal analysis, abnormal movement detection, and remote structural monitoring.
    - Most prominent applications included automatic detection of rockfalls and gas accumulation, as well as confined-space surveillance without human presence.
    - Strategic goals included minimizing worker exposure to high-risk environments and reducing the need for physical inspections.

- Calidus / Seitech with Circet (Spain): AI for Field Safety Monitoring As detailed in section 3.2 on real-time monitoring, Seitech developed *Calidus*, a technological suite aimed at preventive digitalization. In collaboration with Circet Spain, a pilot project was launched in Seville for field operations in telecommunications (Interempresas, 2022b; Seitech Solutions (eFSO), 2023b).

  Key components:

    - Computer vision for automatic PPE verification.
    - Remote assistance through bidirectional video communication between technicians and supervisors.
    - Automated digital records of field inspections and audits.

  Notable results included enhanced regulatory traceability, reduced unnecessary travel, and a strengthened safety culture in decentralized operations.

Other pilot projects in Europe also merit attention for their experimental value:

- Bouygues Construction (France): Tested biometric sensors for predicting worker fatigue (Chen Cheng et al., 2023; Cordier, 2023; Garimella et al., 2020).
- Ferrovial (Spain): Integrated AI into safe activity planning via 4D BIM models (Pastor García et al., 2020; The Factory School, 2025).
- BAM Infra (Netherlands): Used smart-vision drones for aerial inspections and hazardous area mapping (Rodríguez Elizalde, 2022, 2024a).

These examples show that the application of artificial intelligence in construction is no longer a technological utopia but an operational reality in expansion, delivering measurable outcomes in safety, efficiency, and compliance. Despite ongoing cultural, technical, and economic challenges, the pilot projects and field experiences confirm that—under the right conditions—AI is not only viable but highly effective.

The key lies in coherent integration with site operations, active involvement from all stakeholders, and a clear understanding of AI's role: a supportive, predictive, and non-intrusive tool for enabling smarter, distributed, and more sustainable prevention.

6.6 Challenges for the Implementation of AI in Construction Sites

While the potential of artificial intelligence in occupational risk prevention within the construction sector is considerable, its effective deployment faces a range of technical, cultural, and operational challenges that must be properly identified and addressed from the earliest planning stages.

The following are the main obstacles hindering the implementation of intelligent technologies in construction environments (Luna López et al., 2021; Muñoz Perez et al., 2022; Pampliega, 2019):

1. Hostile physical environments for technology: Construction sites are characterized by environmental conditions such as dust, moisture, vibrations, and extreme temperature fluctuations, all of which affect the reliability and durability of technological devices:
    - Temporary structures hinder the stable installation of sensors or cameras.

   - Heavy machinery generates electromagnetic noise and obstructs data collection.

   Partial solution: Use of ruggedized industrial-grade devices (IP67), autonomous portable systems, and hardware architectures designed to withstand harsh environments.

2. Extreme variability across projects: Each construction project is unique, and its execution varies accordingly. This heterogeneity:
   - Requires recalibration of algorithms for each new scenario.
   - Limits the direct reuse of models trained in previous contexts.
   - Demands site-specific setup, connectivity, and data analysis protocols.

   Partial solution: Development of modular architectures with incremental learning and user-friendly configuration tools.

3. Cultural resistance to technological change: A significant portion of personnel perceives these technologies as:
   - Surveillance mechanisms rather than safety tools.
   - A threat to experiential knowledge.
   - Technically inaccessible or limited by language barriers.

   Partial solution: Implement participatory rollouts supported by hands-on training and clear communication that emphasizes direct benefits for workers.

4. Connectivity and power limitations: Especially relevant in civil works located outside urban areas or in rural settings, leading to:
   - Lack of stable network connections or electrical power, restricting cloud-based solutions.
   - Difficulties in continuous data transmission or maintaining devices.

   Partial solution: Adoption of edge computing, mobile networks (4G/5G), and battery or solar-powered systems.

5. Fragmentation of responsibilities and coordination across companies: The presence of multiple subcontractors on the same site causes:
   - Uneven levels of digitalization and incompatible systems.

- Limited integration of safety information.
- Ambiguities regarding system maintenance and accountability.

Partial solution: Deployment of shared platforms with tiered access and clearly defined contractual responsibilities.

6. High initial costs and delayed ROI: Investments in AI involve hardware, training, and support, while returns—such as accident reduction and operational efficiency—may only become evident mid-term.

   Partial solution: Progressive adoption through small-scale pilot projects, evaluation of ROI based on project typology, and leveraging public grants and funding opportunities.

In summary, the failure to implement artificial intelligence in the construction sector does not stem from a lack of technical solutions, but from a lack of strategic integration into the organization and its safety culture. Overcoming these challenges requires shared vision, realistic planning, and a multi-stakeholder commitment. Only then can technology become a true ally in building a modern, effective, and people-centered approach to occupational risk prevention.

6.7 Recommendations for Successful Adoption

The integration of artificial intelligence into the occupational risk prevention landscape in construction requires more than sophisticated tools; it demands a strategic, participatory, and sustainable approach that aligns technology with organizational structure and workplace culture.

The following are key recommendations for effective implementation (Luna López et al., 2021; Muñoz Perez et al., 2022; Pampliega, 2019):

1. Evidence-Based and Gradual Adoption
   - Start with controlled, representative pilot projects.
   - Include key performance indicators from the outset (e.g., incident reduction, system acceptance).
2. Active and Multistakeholder Participation
   - Involve OSH professionals, field workers, supervisors, and technology providers from the design stage.
   - Incorporate formal feedback channels (surveys, focus groups, direct observation).

3. Technical, Functional, and Ethical Training
    - Deliver role-specific training programs (technical, operational, managerial).
    - Address not only system operation but also legal and ethical implications and limitations.
4. User-Centered Design for the Construction Context
    - Interfaces should be intuitive, multilingual, operable with gloves, and functional in noisy environments.
    - Systems should require minimal maintenance, and allow for easy integration and compatibility.
5. Fail-Safe Protocols and Meaningful Human Oversight
    - Define responses to errors, disconnections, or false alarms.
    - Ensure human intervention in all critical decision-making processes.
6. Transparent Internal Communication
    - Clearly explain what the AI system does, how it works, and why it is implemented.
    - Provide information on data processing in accordance with GDPR, fostering trust.
7. Continuous Evaluation and Improvement
    - Conduct technical and organizational audits.
    - Update models and adapt systems to evolving conditions or regulations.
8. Sustainability and Scalability
    - Prioritize modular and interoperable solutions with sustainable financial models.
    - Avoid technological lock-in and plan for medium- and long-term system maintenance.

In short, successful adoption does not depend on deploying the most advanced technology, but on implementing it responsibly, transparently, and with strategic intent (Muñoz Perez et al., 2022). Artificial intelligence can—and should—serve as a catalyst for improving occupational safety in construction, provided it is introduced as a tool for collective empowerment rather than as a mere control mechanism.

7 Case Studies and Real-World Applications of AI in Occupational Risk Prevention

7.1 Importance of Real-World Case Studies

As artificial intelligence gains traction in the construction sector—particularly in the field of occupational risk prevention—a crucial need emerges: to verify that these technologies not only work in theory but are also effective in practice. In this context, real-world case studies represent an indispensable tool.

Concrete examples make it possible to objectively assess the impact of artificial intelligence in active construction environments, involving real workers and dynamic site conditions. Far from being mere anecdotes, these cases provide verifiable evidence, practical lessons, and valuable benchmarks for other organizations in the sector.

To the question of why real-world cases matter, four key responses can be outlined:

1. They validate that technologies function outside the lab. It is not enough for a solution to be technically feasible—it must be fully operational within the actual environment of a construction site, with its inherent risks, uncertainties, and specific pace. Real-world implementations demonstrate this effectiveness in concrete conditions.
2. They show tangible benefits. These include reductions in accident rates, improved compliance with safety standards, cost savings associated with accident prevention, and increases in operational efficiency. Such measurable impacts help justify investment in artificial intelligence from a business and strategic perspective.
3. They facilitate learning from failures. The implementation of any technology is rarely linear. Real-world cases uncover technical, organizational, or human obstacles that other companies can anticipate and navigate more effectively. This shared learning reduces the margin for error and accelerates the adoption curve.
4. They inspire the adoption of similar solutions. Understanding how other organizations—sometimes in comparable contexts—have transformed their preventive culture through artificial intelligence provides both motivation and a framework for action to those taking their initial steps.

The following sections present four representative case studies illustrating how artificial intelligence has been applied in the field of occupational risk prevention in the construction sector. Each example offers a distinct approach, employs varied technologies, and operates in diverse contexts, but all share a common objective: to make work safer through intelligent data use and automation.

In addition to their illustrative value, the cases included in this chapter have been selected based on a series of criteria aimed at ensuring their relevance and educational utility. This is not a random or purely descriptive compilation, but rather a representative and strategic selection of how different AI applications are currently being used—with greater or lesser technological maturity—in real-world occupational settings.

Among the selection criteria applied were the following (Goñi Sein, 2024; Macías García, 2022; Otero Mateo et al., 2023):

- The preventive relevance of the projects—that is, their ability to address key issues related to occupational safety, health, or risk monitoring.
- The degree of technological implementation achieved, prioritizing solutions that are operational or in advanced stages of testing and deployment.
- The functional diversity, aiming to cover different types of technologies (e.g., computer vision, data analytics, automation, robotics) and applications (e.g., detection, training, prediction).
- The accessibility and verifiability of the information, enabling readers to analyze the experiences critically.
- Their replicability value, both technically and organizationally.

These cases should not be interpreted as one-size-fits-all models to be copied indiscriminately, but as contextualized examples, each with its own limitations, that may serve as inspiration for new approaches to applying artificial intelligence in occupational risk prevention—always through a responsible and context-aware lens.

7.2 Case 1: *Smartvid.io* – Video Analysis to Identify Risks on Construction Sites

Smartvid.io is a technology company headquartered in Boston that has developed an artificial intelligence platform aimed at reducing risks on construction sites through automated image analysis. Its solution combines computer vision and machine learning to identify unsafe conditions directly from photos and videos captured in the work environment (Brown, 2019).

At the core of the system lies a deep learning algorithm known as "Vinnie", specifically trained to recognize whether workers are correctly using the required personal protective equipment (PPE)—such as helmets, reflective vests, goggles, gloves—as well as to detect potentially hazardous situations like material accumulation, loose cables, improperly placed ladders, or puddles in walkways (Kanner, 2020).

Each finding is automatically tagged and stored in a database that generates a daily safety score for each jobsite. This index functions as a predictive indicator of the risk level, enabling early identification of accident-prone scenarios and prioritization of corrective measures (Baduge et al., 2022). The greater the number of safety violations detected by the AI system, the higher the risk score assigned for the day.

In the example shown in Figure 19, the interface highlights in red a specific safety breach (e.g., "No Hard Hat"), allowing for immediate intervention by the safety team (Hedmond, 2019).

The added value of this tool lies in its ability to automatically process thousands of images per day, providing visual traceability, systematic detection of unsafe patterns, and remote visibility for safety supervisors. Rather than replacing safety personnel, Smartvid.io acts as a virtual inspector that expands operational reach, allowing action before accidents occur.

This case illustrates how artificial intelligence can transform a traditionally manual practice—visual inspection—into a continuous, objective, and predictive tool, enhancing both surveillance effectiveness and the organization's preventive foresight.

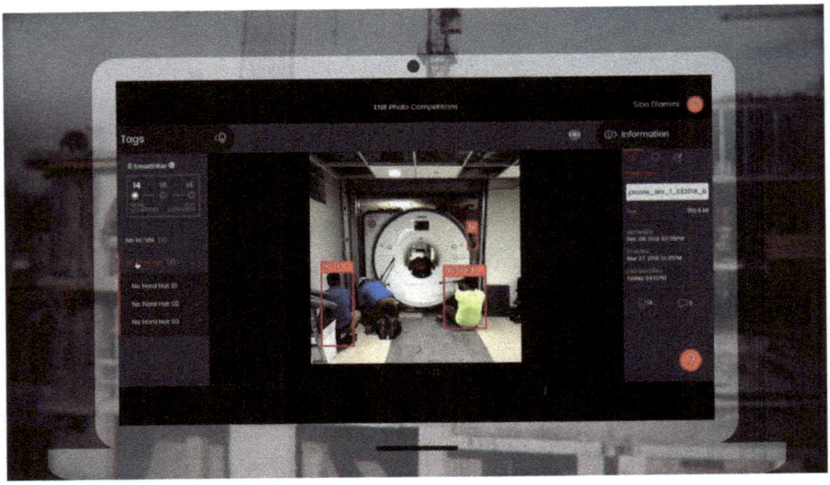

Fig. 19. Smartvid.io interface indicating safety violations (courtesy of Smartvid.io, Hedmond, 2019).

7.2.1 Functionality

The Smartvid.io platform is designed to integrate seamlessly with existing construction management systems, such as Procore, Autodesk BIM 360, or other tools commonly used on-site. Through this integration, the platform automatically collects the images that personnel routinely take with their mobile devices (smartphones, tablets) during visits, inspections, or progress documentation (Hedmond, 2019; Kanner, 2020).

In the background, the "Vinnie" algorithm analyzes these images using computer vision, identifying people, objects, and environmental elements. Its main function is to verify compliance with visible safety conditions (Baduge et al., 2022; Hedmond, 2019): for example, whether workers are wearing helmets, reflective vests, or harnesses correctly; whether unauthorized individuals are in restricted areas; whether equipment is improperly secured; or whether disorder in the workspace could generate risk.

The analysis results are presented on a visual dashboard displaying key metrics such as (Chen et al., 2024; Kanner, 2020):

- PPE compliance rate, expressed as a percentage by type of equipment.
- Number and type of hazardous conditions detected, categorized accordingly.
- Daily global safety score, summarizing the observed risk level.

This dashboard functions both as a real-time monitoring tool and as a decision-support system. Site managers and occupational safety and health (OSH) technicians can receive automated alerts in cases of severe or recurring issues, such as multiple workers at heights without harnesses or critical zones lacking protection (Chen et al., 2024).

The system also features continuous learning: the more visual data it receives, the better it becomes at recognizing new risk scenarios, gradually adapting to the unique characteristics of each site.

Thanks to this design, Smartvid.io not only streamlines safety monitoring, but also improves the predictive and operational capacity of the safety team (Baduge et al., 2022; Kanner, 2020), enabling early interventions, documented reporting, and systematic tracking of unsafe behavior on-site (Garyaev & Garyaev, 2023).

7.2.2 Results

The implementation of Smartvid.io has delivered concrete and verifiable results for various construction companies, establishing its value as a predictive and preventive tool. Notably, Suffolk Construction—one of the leading general contractors in the United States—has achieved significant improvements after integrating this solution into its daily operations.

According to data from one of their pilot initiatives, Smartvid.io's artificial intelligence was able to predict approximately 20% of safety incidents that eventually occurred, with an accuracy rate of 81% (Hedmond, 2019; Kanner, 2020). In other words, four out of five alerts issued by the system corresponded to real situations that resulted—or could have resulted—in accidents.

Applied to a company managing around 50 projects per year, this level of accuracy could, according to internal simulations, help prevent between 40 and 100 incidents annually, resulting in estimated cost savings of $1.4 to $3.6 million, considering disruptions, compensation, litigation, and insurance premiums (Chen et al., 2024; Kanner, 2020).

Recognizing the technology's potential, a consortium of nine major contractors (including Shawmut, DPR, Mortenson, and Bouygues) formed a collaborative data council to enhance the training of the AI model using a broader and more diverse image base. This collaboration has accelerated the algorithm's learning and improved its precision across different types of sites (Baduge et al., 2022).

Beyond the numbers, the impact on daily safety management is significant. In projects using Smartvid.io, there has been a clear reduction in accident rates, driven by a proactive, evidence-based approach (Constance Obiuto et al., 2024). Unsafe trends—such as decreased glove usage, unauthorized area access, or debris accumulation—are addressed immediately during morning briefings and toolbox talks, reinforcing key messages before incidents occur.

For example, if the platform identifies a decline in glove use within a specific crew, the supervisory team can intervene the same day, emphasizing glove use to prevent cuts, abrasions, or other injuries (Constance Obiuto et al., 2024; Rodríguez Elizalde, 2025).

Taken together, the results demonstrate that artificial intelligence can do more than detect existing risks—it can help change unsafe behaviors, enhance preventive oversight, and improve field-level responsiveness.

7.2.3 Benefits

Smartvid.io has proven especially useful as a tool for remote and continuous surveillance. Companies managing multiple construction sites simultaneously use the platform's daily risk index and analyzed imagery to decide where to concentrate their safety personnel most effectively.

During the COVID-19 pandemic, this remote monitoring capability became even more relevant. The platform was updated to automatically detect mask usage and verify physical distancing, thereby supporting compliance with health protocols without requiring inspectors to be physically present on site (Baduge et al., 2022; Chen et al., 2024).

In general, Smartvid.io's artificial intelligence acts as a 24/7 safety assistant, tireless and objective, reviewing all visual materials generated on-site and delivering actionable reports. This not only reduces the number of incidents—by enabling the early detection and elimination of potential hazards—but also improves operational efficiency: OSH technicians spend less time on extensive visual inspections and more time intervening directly in problem areas.

An executive from AECOM emphasized how the implementation of Smartvid.io significantly improved safety metrics in a megaproject in New York, although he also noted that it was necessary to install additional cameras to ensure full coverage of critical work zones (Muñoz Perez et al., 2022; Vélez Aspiazu et al., 2024).

7.2.4 Challenges Faced and Lessons Learned

The implementation of this technology has not been without challenges. One of the first issues identified was the availability and coverage of visual data. For artificial intelligence to detect all present risks, it requires a sufficient volume of representative images for each work area. Some pilot companies reported the need to install additional cameras or to encourage on-site teams to take more daily photographs.

Smartvid.io addressed this challenge by enabling integration with mobile applications already used by operators, thereby avoiding the introduction of additional administrative burdens.

Another significant obstacle was cultural in nature. In the early phases, some workers—upon learning that their images were being analyzed by artificial intelligence—chose to photograph risky situations instead of intervening immediately to correct them. To counter this behavior, companies involved in the pilots implemented training and internal communication actions, reinforcing the idea that technology should be understood as a support for preventive action, not a substitute for human judgment.

Privacy concerns were also addressed. Since the cameras capture images of potentially identifiable individuals, consent and transparent communication mechanisms were established, especially in jurisdictions governed by strict regulations such as the General Data Protection Regulation (GDPR). The platform was also designed to focus exclusively on detecting unsafe conditions and PPE usage, without performing facial recognition or storing personal identities. Data encryption and adherence to international privacy standards have been key in building trust (Garyaev & Garyaev, 2023).

Over time, and following verifiable outcomes—including reductions in accidents, cost savings associated with decreased incident rates, and improvements in safety indicators—Smartvid.io gained acceptance among both executives and frontline workers. It is now consolidated as an integral part of the preventive ecosystem in the companies that have adopted it (Hedmond, 2019; Kanner, 2020).

This case demonstrates how artificial intelligence can be embedded into the daily operational routine of construction sites, delivering real value—as long as it is appropriately contextualized and deployed in alignment with principles of ethics, privacy, and shared responsibility.

7.3 Case 2: *Vinci Construction* – AI-Based PPE Compliance Monitoring

Vinci Construction, one of the world's largest construction groups, headquartered in France, has committed to integrating artificial intelligence into its preventive protocols to strengthen compliance with safety regulations (Constance Obiuto et al., 2024; Mohapatra et al., 2023). Among its various innovation lines, a particularly notable initiative is the implementation of computer vision systems integrated into site surveillance cameras, designed to automatically detect whether workers are properly wearing mandatory personal protective equipment (PPE), such as helmets, reflective vests, and safety footwear (Uribe, 2024b).

Initially deployed as a pilot across several large-scale projects, this technology was developed in collaboration with specialized AI video analytics providers, and it constitutes a key pillar of the company's broader digitalization and preventive safety culture strategy (Boyer et al., 2025).

7.3.1 Functionality

The system developed by Vinci is based on object recognition algorithms trained to automatically identify whether workers are correctly wearing mandatory PPE—such as helmets, reflective vests, or harnesses (Kanner, 2020). It also verifies that only authorized and properly equipped personnel are accessing certain zones.

When a camera identifies a worker in a designated safety area without the required gear, an immediate alert is triggered to the health, safety, and environment (HSE) officer. Depending on the system's configuration, automated actions may follow (Constance Obiuto et al., 2024; Mohapatra et al., 2023):

- Access denial through smart turnstiles.
- Activation of visual or acoustic warning signals.
- Temporary shutdown of dangerous machinery.

These functions are integrated into the site's existing infrastructure, either by adding AI modules to pre-installed CCTV cameras or by installing new units in critical zones (e.g., entrances, loading bays, cranes, or elevated areas). The artificial intelligence processes live video streams, detects safety breaches in real time, and generates both instant alerts and automated statistical reports for further analysis (Boyer et al., 2025).

In addition to PPE detection, the system has demonstrated the ability to identify unsafe behaviors, such as workers walking under suspended loads or approaching operating machinery unsafely (Uribe, 2024b). In such cases, the system may issue voice warnings through loudspeakers, enhancing the capacity for direct, in-situ intervention.

All image analysis is conducted locally, with no storage of personally identifiable data, which ensures compliance with the General Data Protection Regulation (GDPR) by focusing exclusively on visible safety conditions.

7.3.2 Results

The worksites where this system has been deployed have seen an average 25% reduction in safety violations related to PPE usage—especially in secondary access zones and during the early minutes of a shift.

In the medium term, this improved discipline has led to a decrease of up to 40% in the rate of accidents linked to non-compliance with basic safety rules, when compared with previous periods without the support of this technology (Boyer et al., 2025).

Moreover, the system has had a positive training effect: when workers receive immediate alerts for failing to wear the correct equipment, their sense of self-protection and safety culture is reinforced (Christodoulou & Tsoucalas, 2023). This transformation has been accompanied by increased confidence in the system among supervisors and safety managers, who now have access to objective data to support preventive decision-making (Kanner, 2020).

7.3.3 Technology Used

The artificial intelligence system has been modularly integrated into Vinci's digital modeling platform (BIM), enabling the assignment of specific surveillance parameters to each camera location—e.g., detection of face shields in welding areas or verification of harness use in elevated zones.

Detection is based on deep neural networks, such as YOLO or equivalent models, trained on thousands of real-world images from Vinci sites under varied contexts, lighting conditions, and environmental factors (Boyer et al., 2025; Uribe, 2024b). This custom training has been critical for minimizing false positives and increasing system accuracy in demanding conditions.

In some cases, the company evaluated the adoption of commercial off-the-shelf solutions, including localized versions of Smartvid.io, and also collaborated with startups from its own open innovation program (Leonard) to develop custom solutions (Hedmond, 2019; Kanner, 2020).

7.3.4 Challenges Faced and Lessons Learned

One of the main challenges was the cultural acceptance of the technology among on-site personnel. To prevent the system from being perceived as a surveillance tool, Vinci launched an awareness campaign emphasizing that the goal was not punitive, but rather to protect the physical integrity of all workers.

From a technical perspective, integration issues arose with legacy video surveillance systems, prompting the development of hybrid solutions: smart modules added to conventional cameras, followed by a gradual rollout of new AI-native devices.

Regarding regulatory compliance, Vinci configured the system so that all image processing occurs locally, with no transmission or storage of identifying data. Only in cases of repeated violations is a specific protocol triggered that allows safety managers to identify the individual in breach—always within the safeguards established by the GDPR.

A key lesson learned is that trust in the system depends on its reliability: the more accurate and precise the alerts, the faster they are integrated into workers' routines. A phased deployment, coupled with targeted training and measurable results, has enabled Vinci to turn a digital tool into an active component of its safety culture (Christodoulou & Tsoucalas, 2023; Kanner, 2020).

7.4 Case Study 3: Public Works in Japan – AI-Driven Robotics and Sensor Networks for Tunnel Safety

Japan has positioned itself at the global forefront in applying artificial intelligence and robotics to occupational risk prevention, particularly in underground infrastructure projects such as tunnels, dams, and road networks. In response to stringent safety regulations and an increasing shortage of skilled labor, major Japanese construction firms have deployed advanced technological solutions to minimize human exposure to high-risk environments (Mohapatra et al., 2023).

These solutions include autonomous robots for hazardous task execution, intelligent drones for remote inspection, and IoT sensor networks capable of real-time monitoring of structural and environmental conditions inside tunnels (Boyer et al., 2025; Constance Obiuto et al., 2024). As illustrated in Figures 20 and 21, the “Smart Tunnel” system developed by Shimizu Corporation integrates multiple technologies to provide continuous and predictive safety management capabilities (Shimizu Corporation, n.d.-a, n.d.-c).

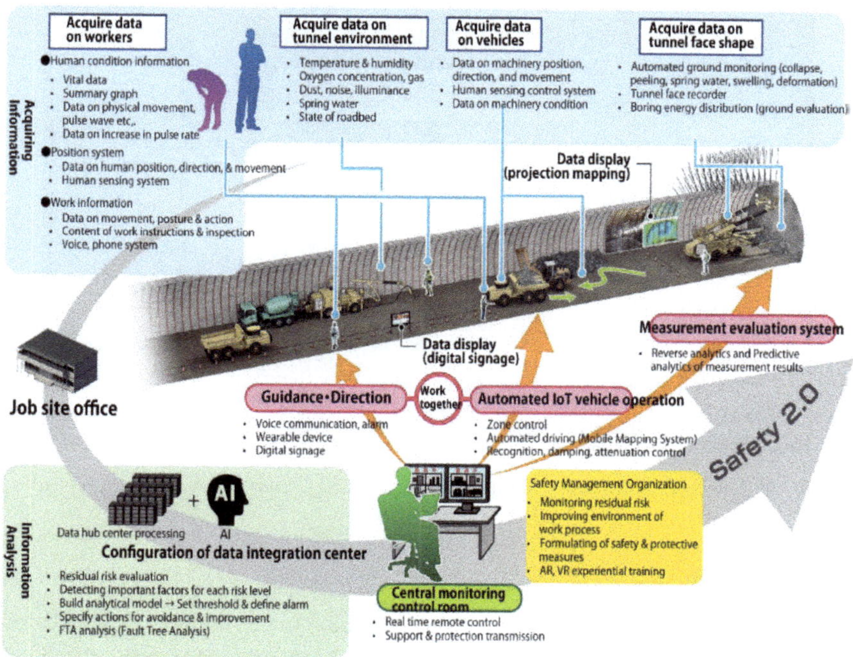

Fig. 20. Diagram of the Smart Tunnel system developed by Shimizu Corporation (Shimizu Corporation, n.d.-b)

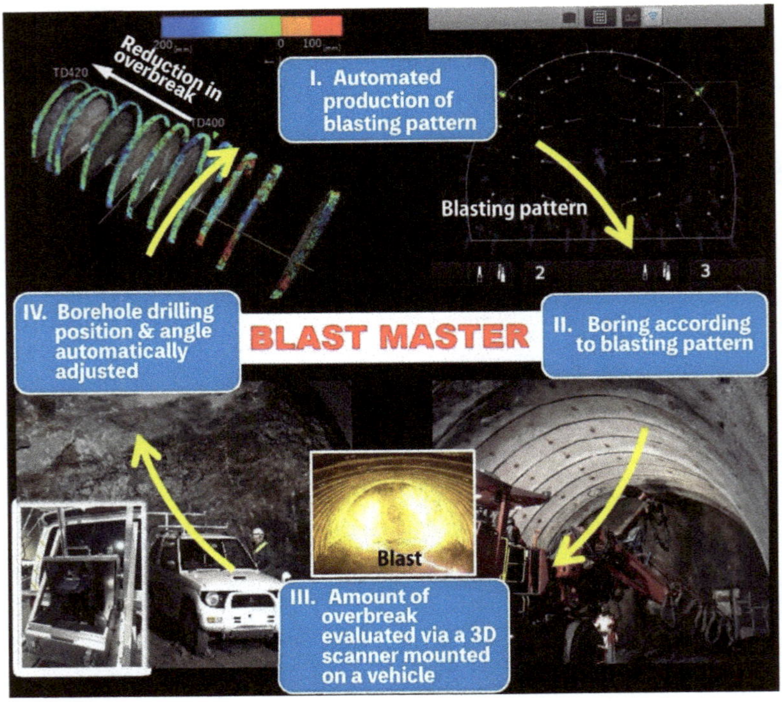

Fig. 21. Conceptual scheme corresponding to Fig. 20 (Shimizu Corporation, n.d.-d).

7.4.1 Operation

One of the most advanced solutions is Shimizu Corporation's Smart Tunnel system (Shimizu Corporation, n.d.-b), which combines AI with distributed sensors to monitor tunnel structural integrity in real time. Devices such as vibration radars installed at excavation fronts (Fig. 22) detect anomalies in the terrain—cracks, shifts, or early warning signs of collapse—and automatically trigger evacuation alerts if safety thresholds are exceeded (McNabb, 2025).

Fig. 22. Intelligent vibration sensor at the excavation front with automated measurement (courtesy of Shimizu Corporation).

This system is enhanced by autonomous drones equipped with computer vision, used post-blasting to inspect freshly excavated tunnel fronts safely and quickly. Instead of dispatching engineers into high-risk zones, the drone navigates through the tunnel, avoids obstacles, and captures high-resolution images. An algorithm analyzes these images to identify fissures, over-excavation, or structural instabilities, and generates near-instant reports.

This system has proven highly effective: post-blast inspections, which traditionally took around two hours, have been reduced to 17 minutes, minimizing risk and accelerating the construction cycle (Rodríguez Elizalde, 2024a, 2024b, 2024c; Toriumi et al., 2023; Zhang et al., 2023).

Additionally, Obayashi Corporation has developed autonomous excavators to remove debris without onboard operators (Bang & Olsson, 2022). Using AI, the system evaluates material distribution and controls the machine arm optimally (Obayashi Corporation, n.d.). Obayashi has also introduced tracked inspection robots capable of entering collapsed areas following earthquakes or landslides, transmitting 3D images in real time to engineers located at safe distances (Mohapatra et al., 2023).

Furthermore, tunnel environments are equipped with IoT environmental sensors that monitor toxic gases, oxygen levels, humidity, temperature, and air quality. These data feed into centralized platforms where anomaly detection algorithms can anticipate critical situations and activate preventative protocols (Shimizu Corporation, n.d.-b; Zhang et al., 2023).

7.4.2 Results

The deployment of these technologies has directly reduced workplace accidents in underground environments. In tunnel segments equipped with vibration radars and systems such as Mimamori Master, no unexpected collapses have been recorded, as any early sign of instability triggers preventative evacuation (Constance Obiuto et al., 2024; Shimizu Corporation, n.d.-d).

Autonomous drones have eliminated the risk associated with post-blasting inspections, and robotic excavators have allowed operations to continue without exposing human operators. In one pilot project, manual inspection was entirely replaced by remote review, reducing incident rates to zero for that task phase.

Moreover, operational efficiency has improved by 10–20% thanks to automation: while a robot inspects one area, other teams can simultaneously prepare the next intervention. This is especially relevant in Japan, where demographic aging and a decline in new entrants into the construction workforce have made generational renewal increasingly difficult (Bang & Olsson, 2022; Mohapatra et al., 2023; Shimizu Corporation, n.d.-b).

7.4.3 Benefits

The integrated approach brings multiple benefits (Constance Obiuto et al., 2024; Shimizu Corporation, n.d.-d):

- Elimination of direct risks for personnel in critical tasks.
- Acceleration of project timelines through task parallelization and automation.
- Improved capacity for early detection of extreme events (earthquakes, gas accumulation, fires).
- Strengthened preventive traceability through continuous digital records.
- Increased confidence among technical teams, who receive accurate, real-time data without entering hazardous zones.

In addition, digital twin technologies have enabled the simulation of complex emergency scenarios and training without putting real lives at risk.

7.4.4 Challenges and Lessons Learned

Despite its success, the implementation of these technologies has faced notable challenges (Shimizu Corporation, n.d.-b, n.d.-d):

- Autonomous navigation in GPS-deprived environments, resolved through Lidar sensors and 3D cameras.
- High initial costs, mitigated by strategic investments from the state and leading construction firms.
- System interoperability, tackled via the adoption of open standards for drones, sensors, and analytics platforms.
- Privacy management, ensured through local data processing and anonymization techniques.
- Personnel reskilling, essential for transitioning traditional roles into new technical functions (e.g., equipment operators becoming remote supervisors, engineers assisted by predictive systems).

In sum, Japan demonstrates that intelligent risk virtualization is not only feasible—it is operational. Machines are deployed first into hostile environments, allowing humans to make decisions based on reliable, real-time data. This is a transformative model of prevention built on technology, anticipation, and human care.

7.5 Caso 4: Construcciones *El Cóndor* (Colombia) – Predictive Risk Platform

Construcciones El Cóndor, one of Colombia's leading infrastructure companies, has made a significant leap in preventive digitalization by developing an artificial intelligence (AI) platform aimed at predicting occupational risks. This solution—designed for roadworks, tunnels, and bridge construction—combines historical data with real-time operational variables to anticipate potential accidents, enabling site teams to make preventive decisions in advance (Jaimes-Quintanilla & Zabala-Vargas, 2024).

7.5.1 Operation

The platform collects data from multiple sources: past incident records, project-specific features, meteorological conditions, operational status, and even indirect indicators such as workforce turnover or estimated workload. Using machine learning algorithms, the system identifies complex correlations and issues predictive alerts based on previously observed patterns (Pampliega, 2019).

For example, under certain conditions (e.g., wet ground, heavy machinery in operation, and newly hired personnel present), the system may detect a high probability of truck overturn and generate a preventive warning, allowing site managers to adjust operations. These alerts are displayed through an interactive dashboard accessible via mobile devices, with risk levels coded by zone and activity type—enabling immediate actions such as reinforcing protective measures, rescheduling tasks, or increasing oversight.

The solution draws not only from internal company data, but also from national statistics provided by Colombia's occupational risk administrator (ARL) (Bongers, 2024). Its modular architecture allows the AI to engage in continuous learning, refining predictions with every new data entry and incorporating key local variables, such as workers' training levels and company-specific practices.

7.5.2 Results

The results have been promising. In projects where the platform has been deployed, there has been a significant reduction in accident rates—particularly in recurring categories such as falls, vehicle overturns, and overexertion. AI-driven alerts have allowed incidents to be avoided by anticipating hazardous scenarios. For instance, during nighttime tasks, the platform identified a heightened risk of ground-level falls due to fatigue and poor lighting, prompting early intervention (Jaimes-Quintanilla & Zabala-Vargas, 2024; Pampliega, 2019).

Beyond accident reduction, the tool has improved operational planning. By forecasting high-risk days, teams are able to proactively reassign work or strengthen preventive measures, leading to greater efficiency and smoother project execution.

7.5.3 Benefits

Construcciones El Cóndor has positioned itself as a regional pioneer in the application of AI for occupational safety and health in Latin America. The platform not only improves predictive capabilities but also fosters a proactive, data-driven safety culture. It is tailored to the Colombian context, incorporating differentiating elements such as geographic variability and the socio-demographic profile of the workforce (Bongers, 2024; Pampliega, 2019).

The model's success has led to plans for scaling the tool, sharing best practices with other construction companies, and strengthening an industry-wide smart prevention network supported by emerging technologies.

7.5.4 Challenges and Lessons Learned

One of the main challenges was the creation of structured, reliable databases. This required digitizing records, correcting inconsistencies, and standardizing data capture across all work units. The key insight: without quality data, effective AI is impossible.

It was also necessary to adapt the algorithmic models to the complexity of risk factors in construction, avoiding misleading correlations and improving system explainability. The active involvement of occupational safety experts was crucial to refine decision rules and align the system with real-world operational conditions (Jaimes-Quintanilla & Zabala-Vargas, 2024).

Finally, managing the cultural shift proved essential. Many mid-level supervisors were initially skeptical of a tool that "suggests" decisions. Through hands-on training and evidence from successful case studies, the platform was gradually adopted as a routine support tool in daily planning.

7.6 Lecciones globales aprendidas

The four case studies presented (Smartvid.io, Vinci Construction, tunnel infrastructure in Japan, and El Cóndor) clearly demonstrate that artificial intelligence (AI) is not an abstract promise, but a real tool capable of transforming occupational risk prevention in the construction sector. However, their success does not depend solely on the technology itself, but on how it is implemented, what data feeds it, how the human factor is considered, and under what organizational and ethical conditions it is deployed. Each case offers valuable insights adaptable to diverse contexts, outlining a roadmap for those aiming to integrate AI into preventive strategies in a realistic, effective, and sustainable manner.

- AI enables a proactive approach to safety. A consistent lesson across the analyzed cases is that AI facilitates a shift from a reactive model—investigating accidents after they occur—to a proactive one that anticipates risks. Solutions such as Smartvid.io and El Cóndor's predictive platform demonstrate how historical data and current conditions can be analyzed to identify risk patterns before they materialize. This has led to significant reductions in incident rates (e.g., up to 40% fewer incidents in Vinci Construction sites after deploying computer vision). The message is clear: investing in predictive prevention saves lives and generates tangible returns by avoiding interruptions, sanctions, and accident-related costs.
- Data quality and volume are critical. The performance of any AI system directly correlates with the quality of its input data. Successful organizations (such as Obayashi, Smartvid.io, and El Cóndor) invested time and resources into collecting, cleaning, and structuring relevant information: site images, historical incident logs, camera footage, IoT sensor readings, and weather data. Poor, incomplete, or mislabelled data can result in false positives, missed risks, or ineffective alerts.

  Thus, a key takeaway is the need to build a robust and updated safety data infrastructure: digitize field reports, deploy appropriate sensors, and standardize protocols for data collection and validation. Cases like Smartvid.io, which improved detection by sharing data among contractors, or Obayashi, with highly digitized construction processes, prove that high-quality data is foundational for smart risk prevention. Keeping this database current ensures AI operates on real, present-day risk patterns.
- Integration with existing systems and workflows. AI should not disruptively overhaul existing processes but rather enhance and integrate

into them. The most successful implementations were those aiming for seamless integration: Smartvid.io connected with familiar platforms like Procore and Autodesk BIM 360; Vinci added smart modules to existing CCTV cameras instead of replacing them; El Cóndor leveraged already-in-use sensors and internal databases to power its predictive system.

Compatibility with existing infrastructures and workflows fosters technical and cultural adoption. Gradual rollouts (starting with pilot projects) allow organizations to validate performance without disrupting construction timelines. Alerts should be delivered through existing communication channels (management software, mobile devices, internal messaging) to avoid being ignored. The key is for AI to support rather than impose.

- Human acceptance and cultural transformation. AI implementation is as much a human challenge as a technological one. Even the most advanced systems are ineffective if not embraced by the organizational culture. All cases (e.g., Vinci, Smartvid.io) noted initial resistance among workers who perceived the systems as surveillance tools or control mechanisms. To overcome this, companies implemented communication and training strategies to clarify that the purpose of AI was protection, not punishment.

  The approach was educational and collaborative: sharing real examples where AI alerts prevented accidents, clearly explaining what data is collected, how it's processed, and the purpose it serves, while assuring that privacy and professional judgment are respected. With visible benefits many workers shifted from skepticism to support.

  Additionally, a data-driven prevention culture was promoted: AI alerts and reports were reviewed in daily briefings, integrated into action plans, and reinforced shared decision-making. Rather than replacing safety staff, AI extends their capabilities: automating routine tasks and freeing time for strategic, on-site prevention.

- Privacy protection and data ethics. The use of AI technologies involving image capture, voice analysis, biometric monitoring, or geolocation requires ethical and legal data management as a top priority. The most advanced organizations in AI-based prevention not only deployed effective systems but also implemented stringent privacy safeguards.

  These measures included video encryption, secure cloud storage with proper certifications, data anonymization or pseudonymization, and strict compliance with data protection laws such as the EU General Data

Protection Regulation (GDPR) or their equivalents elsewhere. Some systems processed data locally (edge computing), avoiding transmission to external servers unless necessary.

Beyond legal compliance, transparency was emphasized to build trust: clearly explaining what data is collected, why, who accesses it, and for how long it's stored. This approach not only prevents legal or labor conflicts but reinforces a company's commitment to technology that enhances safety—not punitive control. In this context, AI ethics is not a rhetorical add-on but a prerequisite for long-term sustainability and legitimacy.

- Top management support and long-term vision. AI-driven safety projects require investment, development time, and patience for system calibration. All successful initiatives had firm backing from top leadership, who recognized safety as a strategic priority. A key lesson is that innovation in occupational safety must be driven from the top, providing resources and breaking the inertia of "we've always done it this way."

  Results may take months to materialize, as AI models require training and data accumulation. Japanese firms, for example, embedded these objectives into long-term strategic plans aiming for full automation over the coming decade. Companies like El Cóndor and Vinci have shown that committed leadership helps solidify a safety policy that goes beyond legal compliance and embraces the transformative power of technology.

  The message is clear: leadership must actively guide change, not merely endorse it. Without executive engagement, AI-based prevention efforts risk becoming isolated experiments with limited impact. With it, prevention becomes a strategic axis and innovation a catalyst for building safer, more efficient, and more humane workplaces.

The deployment of AI, computer vision, robotics, and predictive analytics is tangibly transforming workplace safety in the construction industry. Case studies from the U.S., Europe, Asia, and Latin America show that while approaches vary by context, they all converge on a common goal: to dramatically reduce accident rates.

The lessons throughout this document are unequivocal: acquiring advanced technology is not enough. Intelligent integration into existing workflows, feeding systems with quality data, ensuring technical sustainability, and above all, managing deployment from a human, ethical, and collaborative perspective is essential.

When these elements align, AI becomes not just a future promise but an operational and strategic ally—enhancing preventive effectiveness, reinforcing safety culture, optimizing decision-making, and generating safer, more efficient, and more human-centered work environments.

8 The Future of Artificial Intelligence in Occupational Risk Prevention

8.1 Introduction

Artificial intelligence is no longer a promise of the future—it is now a real and transformative tool within occupational risk prevention systems. Its value is no longer measured solely by its capacity to automate processes, but by its potential to adapt to context, interpret complex signals, and support life-saving decisions that improve working conditions on a daily basis.

This progress is not merely about technology. It represents a profound shift in how we understand and manage safety: moving from reactive, generic models to more anticipatory, personalized, and accountable systems—where people are once again at the center. The future of prevention will not only be digital; it will be smarter, fairer, and above all, more human.

8.2 Emerging Trends

8.2.1 Contextual and Integrated AI

Upcoming AI solutions will no longer function as isolated components but as part of interconnected cyber-physical systems, capable of interpreting the environment across multiple dimensions and in real time.

These systems will combine data from (Cruz-Silva & Gordillo-Pérez, 2022; Goñi Sein, 2024; Nájar Becerra, 2024; Torres Cadillo, 2023):

- The physical environment (temperature, noise, vibration, humidity, etc.),
- The condition of workers (fatigue, stress, cognitive load, heart rate, facial expressions),
- The organizational context (shift schedules, workload, incident history, workplace climate).

This integration will enable systems not only to detect risks, but to understand their root causes and respond proactively, with real-time, personalized adjustments tailored to the task, timing, and worker profile (Goñi Sein, 2024; Torres Cadillo, 2023).

8.2.2 More Accurate Predictive Models

AI algorithms will continue to evolve—not only in terms of processing speed, but in their ability to learn continuously and adapt autonomously.

Technological development will progress along three key lines (Fernández-Costales Muñiz, 2020; Megino Fernández, 2023):

- Anticipating less visible risks, such as psychosocial or ergonomic hazards,
- Learning from real-world workplace dynamics, improving through daily operational feedback,
- Enabling collaborative model training between organizations (federated learning), without centralizing sensitive data—thereby respecting privacy.

This will allow early detection of risk patterns—even before they become evident to experienced human observers (Pineda, 2022).

8.2.3 Natural and Accessible Interfaces

Interaction with intelligent systems will become increasingly simple and intuitive, eliminating the need for complex screens or extensive menus.

Key enabling technologies will include (Beltrán Sosa et al., 2022; Taboada et al., 2022):

- Voice assistants embedded in helmets or smart glasses, providing real-time alerts and guidance,
- Augmented reality (AR) overlays on the physical workspace to guide workers through critical tasks,
- Immersive simulators powered by AI that adapt training content to each worker's profile and pace.

These innovations will not only enhance training effectiveness but also empower workers to actively manage their own safety.

8.3 Evolving Role of OSH Professionals

Digital transformation is reshaping the role of occupational safety and health (OSH) professionals. Their responsibilities will no longer be limited to inspections and standard training delivery.

Emerging OSH profiles will need to:

- Interpret real-time data, validate automated decisions, and detect algorithmic bias,
- Design evidence-based, tech-driven preventive strategies,
- Ensure that the use of AI in the workplace adheres to the principles of fairness, privacy, and human dignity.

Far from diminishing their importance, this evolution places OSH professionals at the center of change, as custodians of the balance between technology, ethics, and safety.

8.4 Evolving Ethical and Regulatory Framework

The use of AI in the workplace raises complex ethical and legal questions (Chicaiza Ortiz, 2024; Lassi, 2023; Torres Cadillo, 2023):

- Is it legitimate to continuously monitor someone at their workstation?
- Who is responsible if an automated alert fails and an accident occurs?
- How can unintentional algorithmic discrimination be prevented?

Regulations such as the forthcoming EU Artificial Intelligence Act will establish minimum standards. However, beyond legal compliance, companies must develop clear, auditable, and ethical internal policies.

Key guiding principles include:

- Algorithmic transparency: Workers should understand how automated decisions are made,
- Right to privacy, especially regarding biometric or behavioral data,
- Shared responsibility between developers, implementers, and system operators.

8.5 Implementation Challenges Ahead

The effective deployment of these technologies will not be without difficulty. Key challenges include:

- The gap between large companies and SMEs, with smaller organizations often lacking resources for implementation,

- Cultural resistance in environments where technology is perceived as a tool of control or threat,
- Digital skills gaps among OSH professionals,
- Rapid technological obsolescence, requiring scalable and continuously updated solutions.

These challenges cannot be solved by investment alone. They demand shared vision, ongoing training, and a participatory culture of change.

8.6 Toward a Digital Preventive Culture

The key to the future lies not only in the tools we use, but in how we integrate them into the way we work.

Artificial intelligence will only be truly effective if it is embedded within a robust safety culture, built on:

- Genuine participation of all professional profiles,
- Shared responsibility between humans and technology,
- Continuous improvement based on data, critical thinking, and operational experience.

Ultimately, the real challenge is not technological—it is cultural: to make innovation a true ally of safe, respectful, and dignified work. Because only human-centered prevention can fully harness the potential of AI without losing its core purpose.

9 General Conclusion

Artificial intelligence has entered the field of occupational risk prevention not merely as a technological advancement, but as a catalyst for a profound transformation in how we understand, assess, and manage workplace safety. This book has followed that process progressively—from the technical foundations of AI to its specific application in the construction sector—integrating ethical, regulatory, and operational perspectives throughout.

First, it has been demonstrated that AI can overcome many of the structural limitations of traditional preventive approaches: reactivity, data scarcity, challenges in dynamic environments, and lack of personalized measures. By automating monitoring, anticipating risk scenarios, and dynamically adapting to real working conditions, AI enhances the preventive capacity of organizations—provided it is used with judgement, oversight, and purpose.

The practical applications analyzed—ranging from real-time monitoring to predictive models, computer vision, virtual assistants, and fatigue detection—make it clear that this is no longer a hypothetical horizon. It is operational reality. However, the mere implementation of advanced solutions is not enough. They must be deployed with a preventive mindset, not just a productive one; with a human-centered approach, not a purely technocratic one.

Artificial intelligence can save lives, but it can also generate new risks: overdependence, erosion of human oversight, privacy violations, reinforcement of bias, or a sense of distrust among staff. Prevention, by its very nature, cannot afford the application of technologies without critical evaluation, ethical reflection, and a clear governance framework. That is why this book has addressed not only the opportunities but also the limits and challenges that come with deploying AI in the workplace.

In the specific case of the construction sector—one of the most complex, variable, and high-risk environments—the integration of AI is emerging as a strategic opportunity. Extreme environmental conditions, high turnover, task concurrency, and the difficulty of on-site supervision mean that technologies such as computer vision, connected sensors, predictive systems, and digital assistants are not luxuries, but necessities. However, their effectiveness will ultimately depend on how well they are integrated with technical expertise, worker participation, and a genuine commitment to safety.

From an organizational perspective, this new scenario demands more than technological investment. It requires new skills (both technical and ethical), flexible structures, data culture, preventive leadership, and responsible innovation. It calls for a redefinition of the role of occupational safety professionals—as analysts, intelligent system designers, and guardians of the integrity of the prevention process.

Therefore, the final recommendations are clear:

- To safety professionals: Acquire competencies in AI not as programmers, but as risk managers of algorithmic systems, critical interpreters of automation, and defenders of technical transparency.
- To companies: Integrate AI as a tool to strengthen the prevention system, not to diminish or outsource responsibilities; and ensure ethical governance of data and algorithms.
- To technology developers: Understand the human and regulatory value of occupational risk prevention, and design human-centered, auditable, explainable, and context-adaptable solutions.
- To lawmakers and public authorities: Support this transformation with a clear and up-to-date regulatory framework, consistent with the principles of fairness, protection, and transparency that underpin labor law.

In short, the future of occupational risk prevention will undoubtedly be more digital, but it will not necessarily be more human. That will depend on our decisions. Because technology alone does not prevent accidents. People—and the organizations that empower them—decide whether to use AI to monitor or to care, to control or to empower, to optimize costs or to save lives.

What is at stake is not merely a change in tools, but a new contract between technology and labor, between safety and productivity, between innovation and dignity. This book has aimed to be a critical compass for navigating that path with rigor, perspective, and commitment.

Because a safer workplace will not be the one with the most sensors—but the one with the most sense.

Appendix I. Regulatory and Legal Framework Applicable to AI in Occupational Risk Prevention

The implementation of artificial intelligence in workplace environments must not—and cannot—be separated from compliance with the legal frameworks that safeguard fundamental rights, health, privacy, and the safety of workers. Far from being an obstacle to innovation, this legal framework provides a guarantee that technology is used in an ethical, transparent, and human-centered manner.

This appendix compiles the main legislative texts and legal references that must be taken into account when applying AI-based solutions in the field of occupational risk prevention, with particular emphasis on the European and Spanish contexts. Understanding and applying these instruments is essential to ensure that the digital transformation of preventive practices remains aligned with the principles of legality, proportionality, and respect for labor rights.

Occupational Safety and Health Legislation

The application of any technology—including artificial intelligence—in the field of occupational safety and health (OSH) must comply with existing legal obligations at national, regional, and international levels. This is not simply about introducing innovation, but about doing so with legal and preventive responsibility, ensuring that workers' rights and safety remain at the core of any technological deployment.

Two key regulatory instruments serve as foundational references:

- European Framework Directive 89/391/EEC on Safety and Health at Work: This directive requires all EU Member States to ensure that employers take necessary measures to safeguard workers' safety and health. It establishes key principles of prevention such as risk avoidance, assessment at source, adapting work to the individual, and systematic planning of prevention strategies. AI-based systems must be aligned with these principles and serve to reinforce—not replace—professional risk assessment and management.
- National OSH Regulations (e.g., Spain's Law 31/1995 on Occupational Risk Prevention): Member States transpose the Framework Directive into national law. These national regulations emphasize that employers retain the non-transferable duty of protection, and that any tool—AI-based or otherwise—must be integrated into formal preventive strategies such as the

risk prevention plan. AI must be assessed as a technical preventive measure, similar to personal protective equipment (PPE) or machinery, with full documentation of its purpose, scope, risks, and control measures.

At the international level, the International Labour Organization (ILO) promotes conventions and recommendations (e.g., Convention No. 155 on Occupational Safety and Health) that reinforce the employer's responsibility to ensure a safe and healthy working environment, regardless of the technology employed. In this context, AI must be implemented in a way that is proportionate, justified, and aligned with preventive objectives.

General Data Protection and Privacy Regulations

The use of artificial intelligence in occupational settings often entails the processing of personal data, particularly in systems that collect images, behavior logs, biometric signals, voice recordings, or other indicators related to workers' health, well-being, or emotional state. This type of processing triggers the direct application of international and regional data protection laws, most notably:

- EU General Data Protection Regulation (GDPR) – Regulation (EU) 2016/679,
- Council of Europe Convention 108+,
- Relevant national data protection acts (e.g., UK Data Protection Act, Brazil's LGPD, etc.).

Under the GDPR—mandatory across the EU and influential globally—AI use in the workplace must comply with several core legal obligations, particularly when automated decision-making or systematic monitoring is involved. Key requirements include:

- Lawful Basis for Processing: Every instance of personal data processing must be supported by a valid legal ground, such as explicit consent, legitimate interest (subject to balancing tests), or legal obligations related to occupational safety and health.
- Transparency and Information: Workers must be clearly and accessibly informed about what data is being collected, for what purpose, how long it will be retained, who will process it, and what safeguards are in place.

- Data Minimization: Only data strictly necessary for the preventive purpose should be processed. The collection or use of excessive, irrelevant, or disproportionate data may constitute a serious infringement.
- Data Protection Impact Assessment (DPIA): Required when the system involves systematic monitoring or automated decision-making with significant effects, such as access restrictions, task assignment changes, or individual alerts. The DPIA must assess risks to rights and freedoms, propose mitigations, and be revisited regularly.
- Guarantee of Digital Rights: Workers retain all rights under the GDPR, including access, rectification, objection, erasure, restriction of processing and notably, the right not to be subject to solely automated decisions that significantly affect their professional status or well-being (Article 22 GDPR).

A relevant example would be an AI system that detects signs of worker fatigue using facial recognition or voice analysis. Such a system must:

- Clearly justify its legal basis.
- Apply data minimization and proportionality principles.
- Use anonymization or pseudonymization techniques whenever possible.
- Offer less invasive alternatives, if available.
- Provide a meaningful option for human review or objection where decisions directly affect individuals.

In summary, data protection is not a mere administrative requirement—it is a structural condition for the legal, ethical, and organizational legitimacy of AI in occupational risk prevention.

European Artificial Intelligence Regulation (*AI Act*)

The European Artificial Intelligence Regulation, commonly referred to as the AI Act, is the world's first comprehensive legislative framework governing the development, use, and oversight of artificial intelligence systems. Following its provisional adoption in 2024, the regulation is expected to enter into force in 2025, with a phased implementation timeline based on levels of risk.

The AI Act introduces a risk-based approach, classifying AI systems according to their potential impact on fundamental rights, health, and human safety. In the field of occupational risk prevention, the regulation identifies the workplace as a particularly sensitive environment.

Relevant Classification for Occupational Risk Prevention

The AI Act establishes several risk categories. In the context of occupational health and safety, the following classifications are particularly relevant:

- High-Risk AI Systems, subject to strict regulatory requirements:
    - Systems that assess workers' behavior, performance, or health status.
    - Applications that automate decisions on task allocation, break scheduling, or operational restrictions.
    - Continuous monitoring tools (e.g., computer vision, biometrics, motion sensors) that may affect workers' autonomy or dignity.
- Limited-Risk AI Systems, which require transparency obligations but are not subject to prior authorization:
    - Virtual assistants used in safety training.
    - Content recommendation systems or good-practice advisories in occupational health and safety.

Requirements for High-Risk Systems

All AI systems classified as high-risk must comply with the following regulatory obligations:

- Meaningful human oversight, with clear mechanisms for control, validation, and intervention in automated decision-making processes.
- Explainability and traceability, ensuring that workers and safety professionals can understand how system decisions are made.
- Documented logging of incidents and corrective measures, as part of the system's operational lifecycle.
- Ethical, inclusive, and bias-free design, both in terms of datasets and algorithmic logic.

These requirements are particularly critical in the domain of occupational risk prevention, where any tool affecting workers' health, safety, or rights must demonstrate compliance with the legal framework and operate under principles of transparency, proportionality, and fairness.

Final Remarks

The entry into force of the AI Act will represent a turning point in the legal management of artificial intelligence applied to occupational safety. Organizations will need to clearly determine whether their systems fall under the high-risk category, and if so, ensure full alignment with the technical, legal, and ethical standards set by the regulation.

This is not merely a matter of regulatory compliance—it is a prerequisite for ensuring that technological innovation in workplace safety is responsible, sustainable, and legitimate.

Otras normas complementarias a considerar

- Ley Orgánica 3/2018 (España) de Protección de Datos y Garantía de Derechos Digitales. Complementa el GDPR con principios específicos sobre el uso de tecnologías en el entorno laboral (artículos 87 a 91), especialmente la obligación de garantizar la dignidad digital en el trabajo.
- Ley 10/2021, de trabajo a distancia. Incluye disposiciones sobre el uso de herramientas tecnológicas para la supervisión y control del rendimiento de personas teletrabajadoras. Es aplicable a la inteligencia artificial empleada para monitorización remota.
- Normas técnicas ISO/IEC
    - ISO/IEC 23894:2023 – Gestión del riesgo en sistemas de inteligencia artificial.
    - ISO 45001 – Sistemas de gestión de la seguridad y salud en el trabajo (incluye aspectos tecnológicos).

Recommendations for Regulatory Compliance

To ensure that the use of artificial intelligence in occupational risk prevention aligns with current legal, ethical, and organizational requirements, the following best practices are recommended:

1. Integrate AI into formal preventive planning, evaluating it as a technical preventive system—just like machinery, procedures, or personal protective equipment. Its function, scope of application, associated risks, and control measures should be documented accordingly.
2. Conduct a legal and ethical impact assessment (LEIA) prior to deployment, particularly if the system involves automated decision-making or systematic monitoring of workers. This assessment should address not only data protection issues, but also proportionality, transparency, and social acceptability.
3. Train occupational health and safety professionals in the legal foundations of data protection, responsible use of intelligent technologies, and the limits of automation in the workplace. The objective is not to turn them into legal experts, but into professionals capable of identifying relevant regulatory and ethical risks.
4. Establish a documented system of meaningful human oversight for all decisions that directly impact individuals (e.g., access control, task assignment, alerts, disciplinary actions), ensuring that human technical judgment is not replaced—but rather reinforced—by automation.
5. Involve the Data Protection Officer (DPO) and the occupational health and safety committee from the early stages of AI system design, procurement, or implementation. Their participation ensures that technological innovation aligns with the principles of responsible governance and active stakeholder involvement.

Conclusion

The legal framework applicable to the use of artificial intelligence in occupational risk prevention is not merely a set of restrictions; it serves as a roadmap for the responsible, legitimate, and sustainable deployment of these technologies. Compliance with regulations should not be seen as an obstacle, but rather as a guarantee of trust, fairness, and effectiveness.

Ultimately, it is not enough for an AI solution to be technically useful—it must also be legally valid, ethically acceptable, and socially legitimate. Its integration into an organization's preventive culture must be carried out with full awareness of its impact on labor rights, individual privacy, and the overall quality of the work environment. Only then can artificial intelligence become a true ally in promoting safe and dignified work.

Appendix II. Glossary of Terms

AI (Artificial Intelligence): A field of computer science that develops systems capable of performing tasks that typically require human intelligence.

Algorithm: A set of instructions followed by a computer to perform a specific task. In artificial intelligence, algorithms are the foundation of learning and automated decision-making.

Artificial Neural Network: A computational model inspired by the human brain, which processes information through interconnected structures (artificial neurons).

Big Data: A collection of technologies and methodologies designed to manage and analyze massive amounts of data that exceed the capacity of traditional systems.

Black Box: A term used to describe AI systems whose internal processes are not easily understandable by humans.

Computer Vision: Technology that enables machines to interpret and process visual information such as images or videos.

Deep Learning: A subfield of machine learning that uses artificial neural networks with multiple layers to process large volumes of complex data.

Digital Twin: A virtual replica of a real-world object, system, or process that allows for simulation and behavior prediction in a controlled environment.

Expert System: A software application that emulates the knowledge of a human expert in a specific domain, providing recommendations or diagnoses.

Internet of Things (IoT): The interconnection of physical objects to the internet, allowing for real-time data collection and sharing.

Machine Learning: A branch of artificial intelligence that enables systems to learn from data without being explicitly programmed.

Natural Language Processing (NLP): A branch of artificial intelligence that enables machines to understand, interpret, and generate human language.

Occupational Risk: The possibility of a worker suffering harm as a result of their job.

PPE (Personal Protective Equipment): A set of devices or garments designed to protect workers from one or more risks that may threaten their safety or health.

Predictive: In the context of prevention, refers to the ability to anticipate an event or risk before it occurs, based on data analysis.

Real-Time Monitoring: Continuous supervision of processes or conditions via automated systems, with the capability to generate alerts or intervene automatically.

Smart Sensors: Devices that capture environmental information and transmit it to systems that interpret it, often using artificial intelligence.

Appendix III. Bibliography and References

Alastruey, C. F. (2021). State of the Art in Artificial Intelligence and Learning Machines. Sociologia y Tecnociencia, 11, 182–195. https://doi.org/10.24197/st.Extra_2.2021.182-195

Alawamleh, M., Shammas, N., Alawamleh, K., & Bani Ismail, L. (2024). Examining the limitations of AI in business and the need for human insights using Interpretive Structural Modelling. Journal of Open Innovation: Technology, Market, and Complexity, 10(3). https://doi.org/10.1016/j.joitmc.2024.100338

Alcobilla Ferrara, E. (2023). Introduction to machine learning in breast disorders. Revista de Senologia y Patologia Mamaria, 36(4). https://doi.org/10.1016/j.senol.2023.100503

Anagnostopoulos, C. N., & Krinidis, S. (2025). Sensors and Advanced Sensing Techniques for Computer Vision Applications. Sensors, 25(1). https://doi.org /10.3390/s25010035

Antezana Soria Galvarro, M. L. (2023). Facing the Challenges of Artificial Intelligence: Ethics, Transparency and the Future. Gaceta Medica Boliviana, 46(2), 5–6. https://doi.org/10.47993/gmb.v46i2.797

Antúnez Cid, J. (2024). Nuestro puesto está en el porvenir. Cruzar el límite del choque de paradigmas: el poder sobre el poder, una reflexión ante el reto de la IA. Revista Espanola de Teologia, 84(3), 391–454. https://doi.org/https://doi.org/10.60098/ret.24084.30001

Aránguiz Villagrán, M. (2022). Auditoría algorítmica para sistemas de toma o soporte de decisiones. https://doi.org/http://dx.doi.org/10.18235/0004154

Ashida, K., Hino, Y., Koopipat, C., Ogawa-Ochiai, K., & Tsumura, N. (2024). Monitoring respiratory state from near-infrared face video images. Artificial Life and Robotics, 29(1), 197–203. https://doi.org/10.1007/s10015-023-00926-3

Baduge, S. K., Thilakarathna, S., Perera, J. S., Arashpour, M., Sharafi, P., Teodosio, B., Shringi, A., & Mendis, P. (2022). Artificial intelligence and smart vision for building and construction 4.0: Machine and deep learning methods and applications. Automation in Construction, 141. https://doi.org/10.1016/j.autcon.2022.104440

Bang, S., & Olsson, N. (2022). Artificial Intelligence in Construction Projects: A Systematic Scoping Review. Journal of Engineering, Project, and Production Management, 12(3), 224–238. https://doi.org/10.32738/JEPPM-2022-0021

Bastidas, M., Jaramillo, P., Chejne, F., & Galván, B. (2010). Análisis multiobjetivo a un sistema energético. Revista Facultad de Ingenieria, 51, 44–51. https://doi.org/10.17533/udea.redin.14919

Beltrán Sosa, L., Luviano Cruz, D., Pérez-Domínguez, L., & Rodríguez Ramírez, A. G. (2022). Realidad aumentada como ayuda visual para una estación de ensamble de manufactura usando lentes inteligentes. Científica, 26(2), 1–8. https://doi.org/10.46842/ipn.cien.v26n2a01

Blasco Jover, C. (2024). IA: dentro de la flexibilidad laboral y de un balance trabajo-vida. Revista Justicia & Trabajo N.a Junio 2024, 2024(4), 77–107. https://doi.org/10.69592/2952-1955-n4-junio-2024-art-3

Bohannon, J. (2015). Fears of an AI pioneer: Stuart Russell argues that AI is as dangerous as nuclear weapons. Science, 349(6245), 252. https://doi.org/10.1126/science.349.6245.252

Boligán Rojas, G., Trinchet Varela, C., & Pérez Rodríguez, R. (2024). Optimización multiobjetivo mediante algoritmos evolutivos, para el diseño de desalinizadores solares de humidificación deshumidificación. Desarrollo Sustentable, Negocios, Emprendimiento y Educación, 6(53), 53–67. https://doi.org/10.51896/rilcods.v6i53.475

Bonfante, F., Maggiore, P., Grimaccia, F., Filippone, E., & Dalla Vedova, M. D. L. (2021). Specific category operations within the u-space: expert systems embedded within rpas as an original mean of mitigation of operational safety risks. International Review of Aerospace Engineering, 14(3), 121–130. https://doi.org/10.15866/irease.v14i3.19985

Bongers, W. (2024). Máquina Cóndor, de Demian Schopf, y los proyectos IA de Sofía Crespo: la imaginación artificial y el desafío noopolítico. Bulletin of Hispanic Studies, 101(6), 579–596. https://doi.org/10.3828/bhs.2024.42

Boyer, C., Guillet, J., Dumas, T., & Oviedo, D. (2025). Management of excavated soils in Co6&Co7 TELT: Addressing valorization challenges and compliance with environmental reporting standards. In Tunnelling into a Sustainable Future–Methods and Technologies (pp. 4511–4518). https://doi.org/https://doi.org/10.1201/9781003559047-574

Brown, K. (2019). By predicting risk, AI tool shapes top firms' safety conversations. Construction Dive. https://www.constructiondive.com/news/by-predicting-risk-ai-tool-shapes-top-firms-safety-conversations/554754/

Caisa Herrera, C. A., Paredes Anchatipán, A. D., & Romero Bedón, F. R. (2024). Análisis del uso de la inteligencia artificial en la toma de decisiones en sistemas de control eléctrico industrial. Revista Científica Multidisciplinar G-Nerando, 5(2). https://doi.org/10.60100/rcmg.v5i2.261

Calvente Menéndez, J. (1996). Prevención de riesgos laborales (Ley 31/1995): las responsabilidades administrativas y penales en materia de seguridad e higiene. Infracciones y sanciones. Revista de Trabajo y Seguridad Social. CEF, 57–96. https://doi.org/10.51302/rtss.1996.18191

Campero-Jurado, I., Márquez-Sánchez, S., Quintanar-Gómez, J., Rodríguez, S., & Corchado, J. M. (2020). Smart helmet 5.0 for industrial internet of things using artificial intelligence. Sensors (Switzerland), 20(21), 1–27. https://doi.org/10.3390/s20216241

Carabantes López, M. (2013). Inteligencia artificial: condiciones de posibilidad técnicas y sociales para la creación de máquinas pensantes [Universidad Complutense de Madrid]. https://docta.ucm.es/entities/publication/0f1d2edb-42c8-44c3-876d-fb966c47b96a

Carrera, P. (2024). Singularity? Limitations, possibilities, and differences of artificial intelligence compared to human intelligence. Claridades, 16(2), 159–186. https://doi.org/10.24310/crf.16.2.2024.19654

Castro Castro, R. J., Chiquito Muñiz, J. J., Romero Castro, M. I., & Clavel Quintero, Y. (2022). La inteligencia artificial y sus diferencias con los sistemas expertos. Journal TechInnovation, 1(2), 88–96. https://doi.org/10.47230/journal.techinnovation.v1.n2.2022.88-96

Celi-Parraga, R. J., Varela-Tapia, E. A., Acosta-Guzmán, I. L., & Montaño-Pulzara, N. R. (2021). Técnicas de procesamiento de lenguaje natural en la inteligencia artificial conversacional textual. AlfaPublicaciones, 3(4.1), 40–52. https://doi.org/10.33262/ap.v3i4.1.123

Chen Cheng, C., Chung, E., & Correa, N. (2023). inteligencia Artificial y su Impacto en la Industria de la Ingeniería. Reicit, 3(1), 26–40. https://doi.org/10.48204/reict.v3n1.3948

Chen, K., Zhou, X., Bao, Z., Skibniewski, M. J., & Fang, W. (2024). Artificial intelligence in infrastructure construction: A critical review. Frontiers of Engineering Management. https://doi.org/10.1007/s42524-024-3128-5

Chernyshev, D., Dolhopolov, S., Honcharenko, T., Haman, H., Ivanova, T., & Zinchenko, M. (2022). Integration of Building Information Modeling and Artificial Intelligence Systems to Create a Digital Twin of the Construction Site. International Scientific and Technical Conference on Computer Sciences and Information Technologies, 2022-Novem, 36–39. https://doi.org/10.1109/CSIT56902.2022.10000717

Chicaiza Ortiz, W. (2024). La inteligencia artificial en auditoría: riesgos éticos y requisitos normativos. ECiencia, 1(3), 31–44. https://doi.org/https://doi.org/10.71022/jtl1iy06

Christodoulou, K., & Tsoucalas, G. (2023). Artificial Intelligence: From Talos to da Vinci. European Journal of Therapeutics, 29(3), e25–e27. https://doi.org/10.58600/eurjther1775

Cogliati, D., Falchetto, M., Pau, D., Roveri, M., & Viscardi, G. (2018). Intelligent cyber-physical systems for industry 4.0. Proceedings - 2018 1st IEEE International Conference on Artificial Intelligence for Industries, AI4I 2018, 19–22. https://doi.org/10.1109/AI4I.2018.8665681

Constance Obiuto, N., Adekola Adebayo, R., Kayode Olajiga, O., & Clinton Festus-Ikhuoria, I. (2024). Integrating Artificial Intelligence in Construction Management: Improving Project Efficiency and Cost-effectiveness. International Journal of Advanced Multidisciplinary Research and Studies, 4(2), 639–647. https://doi.org/10.62225/2583049x.2024.4.2.2550

Cordier, J. (2023). Quelle est la valeur ajoutée de l'IA pour les pros de la construction, l'immobilier et l'urbanisme ? Bouygues Construction. https://www.bouygues-construction.com/blog/fr/ia-construction-immobilier-urbanisme/

Corvalán, J. G. (2018). Inteligencia artificial: Retos, desafíos y oportunidades - Prometea: La primera inteligencia artificial de Latinoamérica al servicio de la Justicia. Revista de Investigacoes Constitucionais, 5(1), 295–316. https://doi.org/10.5380/rinc.v5i1.55334

Cruz-Silva, J., & Gordillo-Pérez, S. (2022). Inteligencia artificial en el campo laboral: conflicto de rol y bienestar. Redmarka. Revista de Marketing Aplicado, 26(1), 52–61. https://doi.org/10.17979/redma.2022.26.1.9041

De Paula, G. (2023). Toma de decisiones, inteligencia criminal y nuevas herramientas de análisis: el uso de la inteligencia artificial. Revista de La Escuela Nacional de Inteligencia, 1(2), 133–149. https://doi.org/10.58752/azfvq42s

Del Val y de la Fuente, E. (1996). Prevención de riesgos laborales (Ley 31/1995): derechos y obligaciones de las empresas y de los trabajadores. Revista de Trabajo y Seguridad Social. CEF, 3–28. https://doi.org/10.51302/rtss.1996.18183

Delhi, V. S. K., Sankarlal, R., & Thomas, A. (2020). Detection of Personal Protective Equipment (PPE) Compliance on Construction Site Using Computer Vision Based Deep Learning Techniques. Frontiers in Built Environment, 6. https://doi.org/10.3389/fbuil.2020.00136

Diestra Quinto, N. M., Cordova Villodas, A. J., Caruajulca Montero, C. P., Esquivel Cueva, D. L., & Nina Vera, S. A. (2021). La inteligencia artificial y la toma de decisiones gerenciales. Revista de Investigación Valor Agregado, 8(1), 52–69. https://doi.org/10.17162/riva.v8i1.1631

Dolhopolov, S., Honcharenko, T., Savenko, V., Balina, O., Bezklubenko, I. & Liashchenko, T. (2023). Construction Site Modeling Objects Using Artificial Intelligence and BIM Technology: A Multi-Stage Approach. SIST 2023 - 2023 IEEE International Conference on Smart Information Systems and Technologies, Proceedings, 174–179. https://doi.org/10.1109/SIST58284.2023.10223543

Ekmekci, P. E., & Arda, B. (2020). History of Artificial Intelligence. SpringerBriefs in Ethics, 1–15. https://doi.org/10.1007/978-3-030-52448-7_1

Estardid Colom, F. (2023). La formación en prevención de riesgos laborales ante el impacto tecnológico. Lan Harremanak - Revista de Relaciones Laborales, 49, 89–114. https://doi.org/10.1387/lan-harremanak.24843

European Agency for Safety and Health at Work (EU-OSHA) (2022). Occupational safety and health in the construction sector – Overview. Occupational Safety and Health in the Construction Sector – Overview. https://osha.europa.eu/en/themes/construction

EU-OSHA. (2018). Cultura Preventiva - Safety and health at work EU-OSHA. Cultura Preventiva - Safety and Health at Work EU-OSHA. https://osha.europa.eu/es/tools-and-resources/eu-osha-thesaurus/term/62012d

Fernández-Costales Muñiz, J. (2020). La prevención de riesgos laborales y sus nuevas exigencias y retos frente al avance de la digitalización y las nuevas tecnologías. Revista de Trabajo y Seguridad Social. CEF, 83–115. https://doi.org/10.51302/rtss.2020.934

Ficapal Cusí, P. (2024). IA, automatización y trabajo humano: de la carrera al entendimiento. Oikonomics, 23. https://doi.org/https://doi.org/10.7238/o.n23.2418

Freré Arauz, J. S., Véliz Gavilanes, J. P., Sarco Alemán, E. M., & Campoverde Jimenez, K. J. (2022). La percepción, la cognición y la interactividad. Recimundo, 6(2), 151–159. https://doi.org/10.26820/recimundo/6.(2).abr.2022.151-159

Galván Guerra, R., Cantera Cantera, L. A., Velázquez Velázquez, J. E., Lozano Hernández, Y., & Villafuerte Segura, R. (2024). Ecosistemas Inteligentes. Pädi Boletín Científico De Ciencias Básicas E Ingenierías Del ICBI, 12(Especial 4). https://doi.org/https://doi.org/10.29057/icbi.v12iEspecial4.14193

Garimella, S. A., Senouci, A., & Kim, K. (2020). Monitoring fatigue in construction workers using wearable sensors. Construction Research Congress 2020: Safety, Workforce, and Education - Selected Papers from the Construction Research Congress 2020, 86–94. https://doi.org/10.1061/9780784482872.010

Garyaev, A., & Garyaev, N. (2023). Integration of artificial intelligence and video surveillance technology to monitor construction equipment. E3S Web of Conferences, 410. https://doi.org/10.1051/e3sconf/202341004002

Goñi Sein, J. L. (2024). Sistemas de inteligencia artificial y prevención de los riesgos laborales. Obligaciones del proveedor y del empresario. Labos, 5, 154–484. https://doi.org/https://doi.org/10.20318/labos.2024.9036

González Guerrero, E. (2021). Algoritmos, poder y responsabilidad del ingeniero. Revista SISTEMAS, 161, 4–7. https://doi.org/10.29236/sistemas.n161a1

Greenemeier, L. (2018). Uber Self-Driving Car Fatality Reveals the Technology's Blind Spots. Scientific American, 21. https://www.scientificamerican.com/article/uber-self-driving-car-fatality-reveals-the-technologys-blind-spots1/

Gutierrez-Vasques, X., & Mijangos de la Cruz, V. G. (2024). De las ideas verdes incoloras hasta ChatGpt: los grandes modelos del lenguaje. TIES, Revista de Tecnología e Innovación En Educación Superior, 10, 12–23. https://doi.org/10.22201/dgtic.26832968e.2024.10.18

Hedmond, S. (2019). Smartvid.io Uses Artificial Intelligence to Predict Safety Risks on the Jobsite. Construction Junkie. https://www.constructionjunkie.com/blog/2019/11/10/smartvidio-uses-artificial-intelligence-to-predict-safety-risks-on-the-jobsite

Instituto Nacional de Seguridad y Salud en el Trabajo. (2024). Informe anual de accidentes de trabajo 2023. Informe Anual de Accidentes de Trabajo 2023. https://www.insst.es/documentacion/material-tecnico/documentos-tecnicos/informe-anual-de-accidentes-de-trabajo-en-espana

Interempresas. (2022a). Calidus automatiza la prevención de riesgos laborales gracias a la visión artificial. Calidus Automatiza La Prevención de Riesgos Laborales Gracias a La Visión Artificial. https://www.interempresas.net/Proteccion-laboral/Articulos/405516-Seitech-lanza-Calidus-automatizar-prevencion-riesgos-laborales-gracias-Vision-Artificial.html

Interempresas. (2022b). Calidus automatiza la prevención de riesgos laborales gracias a la visión artificial. Calidus Automatiza La Prevención de Riesgos Laborales Gracias a La Visión Artificial.

Jaimes-Quintanilla, M., & Zabala-Vargas, S. (2024). Artificial intelligence in project management: case of construction and civil works. European Public and Social Innovation Review, 9. https://doi.org/10.31637/epsir-2024-1615

Jones, A. (2024). How AI is revolutionizing productivity, efficiency and knowledge sharing. Skanska. https://www.usa.skanska.com/who-we-are/media/constructive-thinking/how-ai-is-revolutionizing-productivity-efficiency-and-knowledge-sharing/

Jones, L. D., Golan, D., Hanna, S. A., & Ramachandran, M. (2018). Artificial intelligence, machine learning and the evolution of healthcare: A bright future or cause for concern? Bone and Joint Research, 7(3), 223–225. https://doi.org/10.1302/2046-3758.73.BJR-2017-0147.R1

Kalogianni, E., Floros, G. S., & Dimopoulou, E. (2021). Investigating transport infrastructure objects within their spatial development lifecycle. ISPRS Annals of the Photogrammetry, Remote Sensing and Spatial Information Sciences, 8(4/W2-2021), 129–136. https://doi.org/10.5194/isprs-annals-VIII-4-W2-2021-129-2021

Kanner, J. (2020). Smartvid.io : "our AI, 'Vinnie', can help VINCI reduce the risk of accidents at its sites." Leonard. https://leonard.vinci.com/en/smartvi-io-our-ai-vinnie-can-help-vinci-reduce-the-risk-of-accidents-at-its-sites/

Kawasaki, S., Ashida, K., Nguyen, V. T., Ngo, T. D., Le, D. D., Doi, H., & Tsumura, N. (2024). Intrinsic Motivational States Can Be Classified by Non-Contact Measurement of Autonomic Nervous System Activation and Facial Expressions. Applied Sciences (Switzerland), 14(15). https://doi.org/10.3390/app14156697

Kawasaki, S., Kinefuchi, M., Hino, Y., Kobayashi, A., Kawahito, S., Takahashi, M., & Tsumura, N. (2025). Performance evaluation of blood pressure estimation using an RGB-NIR camera. Artificial Life and Robotics. https://doi.org/10.1007/s10015-025-01015-3

Kuenzel, R., Mueller, M., Teizer, J., & Blickle, A. (2015). SmartSite: Intelligent and autonomous environments, machinery, and processes to realize smart road construction projects. 32nd International Symposium on Automation and Robotics in Construction and Mining: Connected to the Future, Proceedings. https://doi.org/10.22260/isarc2015/0068

Laad, M., Maurya, R., & Saiyed, N. (2024). Unveiling the Vision: A Comprehensive Review of Computer Vision in AI and ML. 2024 International Conference on Advances in Data Engineering and Intelligent Computing Systems, ADICS 2024. https://doi.org/10.1109/ADICS58448.2024.10533631

Landa Arroyo, C. (2021). Constitución, Derechos Fundamentales, Inteligencia Artificial y Algoritmos. THEMIS Revista de Derecho, 79, 37–50. https://doi.org/10.18800/themis.202101.002

Lassi, A. (2023). Ética De Los Algoritmos. Revista Ciencias Sociales, 1(44), 145–177. https://doi.org/10.29166/csociales.v1i44.4213

Leguizamón Páez, M. A., & Becerra Correa, N. (2024). Inteligencia Artificial Pasado, presente y futuro. XIKUA Boletín Científico de La Escuela Superior de Tlahuelilpan, 12(Especial), 81–88. https://doi.org/10.29057/xikua.v12iespecial.12747

León Cuenca, O. C., Caraguay Tandazo, J., & Ruiz Ordóñez, R. E. (2024). Evolución de la Inteligencia Artificial y su impacto en la Educación: Revisión de la literatura. Revista de Investigación Científica TSE DE, 7(1). https://doi.org/10.60100/tsede.v7i1.195

León García, R. (2023). Presentación. La inteligencia artificial. Cuadernos Fronterizos, 58, 28–30. https://doi.org/10.20983/cuadfront.2023.58.8

Llata, J. R., Sarabia, E. G., Fernández, D., Arce, J., & Oria, J. P. (2000). Aplicacion de Inteligencia Artificial en Sistemas Automatizados de Produccion. Inteligencia Artificial, 4(10). https://doi.org/10.4114/ia.v4i10.668

Lopez C., C. A., Ferrin, C. D., & Castillo, L. F. (2018). Visión por Computador aplicada al control de calidad en procesos de manufactura: seguimiento en tiempo real de refrigeradores. Revista EIA, 15(30), 57–71. https://doi.org/10.24050/reia.v15i30.1150

López de Mántaras, R. (2020). El traje nuevo de la inteligencia artificial. Investigación y Ciencia, 526, 52–59.

López Eguiarte, A. I., Ramírez Gallo, D. A., González González, P. S., & López Velarde, E. F. (2024). Modelos GPT como Apoyo Tecnológico para Romper la Brecha Digital. Estudios y Perspectivas Revista Científica y Académica, 4(2), 1767–1786. https://doi.org/10.61384/r.c.a..v4i2.319

Luna López, M., Hernández Lozano, M., Aldana Franco, R., Álvarez Sánchez, E., Leyva Retureta, J. G., Ricaño Herrera, F., & Aldana Franco, F. (2021). Sistema inteligente de monitoreo para condiciones ambientales en Industria 4.0. Científica, 25(2), 1–10. https://doi.org/10.46842//ipn.cien.v25n2a07

Macías García, M. del C. (2022). La inteligencia artificial para el entorno laboral. Un enfoque en la predicción de accidentes. E-Revista Internacional De La Proteccion Social, 7(1), 8–101. https://doi.org/10.12795/e-rips.2022.i01.05

Macías García, M. del C. (2023). La inteligencia artificial. Custodia de la seguridad y salud de las personas trabajadoras. E-Revista Internacional De La Proteccion Social, extra, 219–237. https://doi.org/10.12795/e-rips.2023.mon.13

Márquez-Sánchez, S., Campero-Jurado, I., Herrera-Santos, J., Rodríguez, S., & Corchado, J. M. (2021). Intelligent platform based on smart ppe for safety in workplaces. Sensors, 21(14). https://doi.org/10.3390/s21144652

Massiris, M., Fernández, J. A., Bajo, J., & Delrieux, C. (2021). An automated system for monitoring the use of personal protective equipment in the construction industry. RIAI - Revista Iberoamericana de Automatica e Informatica Industrial, 18(1), 68–74. https://doi.org/10.4995/RIAI.2020.13243

Matías, J. M., Rivas, T., Martín, J. E., & Taboada, J. (2008). A machine learning methodology for the analysis of workplace accidents. International Journal of Computer Mathematics, 85(3–4), 559–578. https://doi.org/10.1080/00207160701297346

Mayo de Com, H. C. (2021). La Visión Por Computadora Y Las Futuras Aplicaciones Tecnológicas En Diversos Escenarios. Revista de La Academia Del Guerra Del Ejército Ecuatoriano, 12(1), 5. https://doi.org/10.24133/age.n12.2019.13

Mazzetto, S. (2024). Integrating Emerging Technologies with Digital Twins for Heritage Building Conservation: An Interdisciplinary Approach with

Expert Insights and Bibliometric Analysis. Heritage, 7(11), 6432–6479. https://doi.org/10.3390/heritage7110300

McNabb, M. (2025). Drones Go Underground: Japan's Toda Construction Uses UAVs to Inspect Tunnel Faces. Drone Life. https://dronelife.com/2025/04/09/drones-go-underground-japans-toda-construction-uses-uavs-to-inspect-tunnel-faces/

Megino Fernández, D. (2023). La formación como principio y fin de la acción preventiva frente a los riesgos psicosociales derivados de la digitalización y la automatización. Lan Harremanak - Revista de Relaciones Laborales, 49, 115–154. https://doi.org/10.1387/lan-harremanak.24829

Mellal, M. A. (2020). Obsolescence – A review of the literature. Technology in Society, 63. https://doi.org/10.1016/j.techsoc.2020.101347

Mendizá, C. A., & Jiménez, N. Z. (2021). Governance of Robots for Their Proper Impact on Society: What Tools Are Available? Arbor, 197(802). https://doi.org/10.3989/arbor.2021.802003

Mir, E. (2023). El cambio de paradigma de la Inteligencia Artificial. EME Experimental Illustration, Art & Design, 11, 64–75. https://doi.org/10.4995/eme.2023.19249

Mohan, R., Roselyn, J. P., & Uthra, R. A. (2023). LSTM based artificial intelligence predictive maintenance technique for availability rate and OEE improvement in a TPM implementing plant through Industry 4.0 transformation. Journal of Quality in Maintenance Engineering, 29(4), 763–798. https://doi.org/10.1108/JQME-07-2022-0041

Mohapatra, A., Mohammed, A. R., & Panda, S. (2023). Role of Artificial Intelligence in the Construction Industry – A Systematic Review. Ijarcce, 12(2). https://doi.org/10.17148/ijarcce.2023.12205

Monereo Pérez, J. L., & López Insua, B. del M. (2016). El derecho a la prevención: Un análisis del modelo organizativo preventivo español. Revista de Trabajo y Seguridad Social. CEF, 17–38. https://doi.org/10.51302/rtss.2016.2024

Montenegro Meza, M. A., Méndez, R. M., & Méndez, R. M. (2023). Una introducción amable pero rigurosa al aprendizaje por refuerzo. Recibe,

Revista Electrónica De Computación, Informática, Biomédica Y Electrónica, 12(1), C1-1-C1-15. https://doi.org/10.32870/recibe.v12i1.268

Morales-Conejo, L. (2021). Uso de tecnologías disruptivas con BIM (Building Information Modelling). Revista Tecnología En Marcha. https://doi.org/10.18845/tm.v34i7.6017

Moreno Rabel, M. D., & Fernández Muñoz, J. Á. (2022). Desarrollo de una herramienta de configuración asistida de Sistemas Automatizados de Prevención de Riesgos Industriales basados en Visión Artificial. 963–970. https://doi.org/10.17979/spudc.9788497498081.0963

Mosquera, R., Parra, L., Ledesma, A. J., & Bonilla, H. F. (2021). Applying data mining techniques to predict occupational accidents in the pulp and paper industry. Informacion Tecnologica, 32(1), 133–142. https://doi.org/10.4067/S0718-07642021000100133

Muñoz Perez, S. P., Mendoza Peña, J. G., & Quispe Vílchez, M. B. (2022). Una revisión sobre el rol de la inteligencia artificial en la industria de la construcción. Ingeniería Y Competitividad, 24(02), 23. https://doi.org/10.25100/iyc.v0i00.11727

Nájar Becerra, C. A. (2024). La Inteligencia Artificial y los riesgos psicosociales en el trabajo: El deber de prevención frente al uso de las nuevas tecnologías en las dinámicas laborales del siglo XXI. Laborem, 22(29), 83–107. https://doi.org/10.56932/laborem.22.29.4

Niebles, J. C. (2020). Inteligencia artificial en todo y para todos. Revista Digital Universitaria, 21(1). https://doi.org/10.22201/codeic.16076079e.2020.v21n1.a5

Obayashi Corporation. (n.d.). ESG Management and Technological Innovation—Paving the Way for a Sustainable Future. Obayashi Chronicle. https://www.obayashi.co.jp/chronicle/130th/en/history/chapter3_10.html#:~:text=use,highly precise level of control

Organización Internacional del Trabajo. (2023). Safety and health at the heart of the future of work: Building on 100 years of experience. Safety and Health at the Heart of the Future of Work: Building on 100 Years of Experience. https://www.ilo.org/global/topics/safety-and-health-at-work/lang--en/index.htm

Ortega Giménez, A., Gonzalo Domenech, J. J., & Bonmatí Sánchez, J. (2021). La aplicación de la inteligencia artificial y el derecho: La gestión de riesgos como fundamento de la diligencia debida frente a los riesgos de la inteligencia artificial. CEFLegal. Revista Práctica de Derecho, 5–36. https://doi.org/10.51302/ceflegal.2021.9467

Otero Mateo, M., Pastor Fernández, A., Cerezo Narváez, A., Castilla Barea, M. & Bastante Ceca, M. J. (2023). Impact of Artificial Intelligence in Industrial Collaborative Environments. Proceedings from the 27th International Congress on Project Management and Engineering. https://doi.org/10.61547/3411

Otero Parga, M. M. (2023). Can Artificial Intelligence Substitute the Human Mind? Implications of AI in Fundamental Rights and Ethics. Anales de La Catedra Francisco Suarez, 57, 39–61. https://doi.org/10.30827/acfs.v57i.24710

Pampliega, C. J. (2019). Inteligencia artificial en el sector de la construcción = Artificial inteligence in construction sector. Building & Management, 3(2), 1. https://doi.org/10.20868/bma.2019.2.3917

Pantoja-Rodríguez, J. P., Vera-Gutiérrez, S. E., & Avilés-Flor, T. Y. (2017). Riesgos laborales en las empresas. Polo Del Conocimiento, 2(5), 833. https://doi.org/10.23857/pc.v2i5.98

Park, K., Seokbeen, I., Sehoon, K., & Kiyoung, K. (2020). A Study on Institutional Improvement for Application ofSmart Construction Technology. Journal of the Korea Institute of Construction Safety, 3(1), 9–14. https://doi.org/https://doi.org/10.20931/JKICS.2020.3.1.009

Pastor García, R., Collado Mariscal, D., & Cuevas Murillo, A. (2020). El papel de BIM en la industria 4.0. Cimbra: Revista Del Colegio de Ingenieros Técnicos de Obras Públicas, 417, 6–14. https://www.yumpu.com/es/document/read/63694751/cimbra-417-especial-digital

Peña-Cáceres, O., Silva-Marchan, H., Espinoza-Nima, R., & More-More, M. (2025). Cyber-physical systems in 21st century education. European Public and Social Innovation Review, 10, 1–19. https://doi.org/10.31637/epsir-2025-636

Pérez-López, E. (2015). Los sistemas SCADA en la automatización industrial. Revista Tecnología En Marcha, 28(4), 3. https://doi.org/10.18845/tm.v28i4.2438

Pérez Herrero, P. (2023). Trabajo, individualismo y mundialización en el siglo XXI. In Trabajo, individualismo y mundialización en el siglo XXI. Marcial Pons. https://doi.org/10.2307/j.ctv37xg1fk

Pineda, J. M. (2022). Predictive models in health based on machine learning. Revista Medica Clinica Las Condes, 33(6), 583–590. https://doi.org/10.1016/j.rmclc.2022.11.002

Pino V., L. E., Rico-Carrillo, A. E., & Hernández-Arango, A. (2022). Del ábaco a las redes neuronales o la breve historia de la inteligencia artificial en salud. Medicina, 43(4), 514–526. https://doi.org/10.56050/01205498.1641

Ponce Solé, J. (2024). Inteligencia Artificial, decisiones administrativas discrecionales totalmente automatizadas y alcance del control judicial: ¿indiferencia, insuficiencia o deferencia? Revista de Derecho Público: Teoría y Método, 9, 171–220. https://doi.org/10.37417/rdp/vol_7_2024_2151

Portillo-Ramírez, G., Cruz-Suárez, H., & Velasco-Luna, F. (2024). Un panorama general del aprendizaje por refuerzo en líneas de espera controladas a tiempo discreto. CIENCIA Ergo-Sum, 32. https://doi.org/10.30878/ces.v32n0a31

Ramírez Véliz, R. B., López Solís, S. C., & Garzón Balcázar, J. M. (2022). El humano y la máquina: perspectivas sobre inteligencia artificial, agentes y sistemas inteligentes. Reciamuc, 6(3), 490–501. https://doi.org/10.26820/reciamuc/6.(3).julio.2022.490-501

Retto-Manrique, J. (2021). Las Claves Del Aprendizaje Artificial. Paideia XXI, 11(1), 191–202. https://doi.org/10.31381/paideia.v11i1.3772

Reyes Cárdenas, P. (2023). ¿Qué tipo de "inteligencia" es la Inteligencia Artificial? Metafísica y Persona, 30, 39–48. https://doi.org/10.24310/metyper.2023.vi30.17375

Rico-Bautista, D., Maestre-Góngora, G. P., Guerrero, C. D., Medina-Cárdenas, Y., Areniz-Arévalo, Y., Sanchez-Velasquez, M. C., & Barrientos-Avendaño, E.

(2021). Smart university: Key factors for the adoption of internet of things and big data. RISTI - Revista Iberica de Sistemas e Tecnologias de Informacao, 2021(41), 63–79. https://doi.org/10.17013/RISTI.41.63-79

Ríos, M. (2023). Inteligencia Artificial: Cuando la tecnología es el menor de los paradigmas (Artificial Intelligence: When Technology Is the Smallest of the Paradigms). SSRN Electronic Journal. https://doi.org/10.2139/ssrn.4521736

Rodríguez-Martínez, A., Pérez-Villar, J., & Martín-Cano, M. del C. (2023). Ética, Responsabilidad Y Uso De Tecnologías En La Intervención Social. Antropología Experimental, 23, 81–94. https://doi.org/10.17561/rae.v23.7843

Rodríguez Elizalde, R. (2022). Structural Inspection by RPAS (Drones): Quality Work with Preventive Guarantee. Journal of Engineering and Applied Sciences Technology, 1–11. https://doi.org/10.47363/jeast/2022(4)143

Rodríguez Elizalde, R. (2024a). Comprehensive Technical Inspection of a Medieval Bridge (Ponte de Vilanova, in Allariz) Using Microtechnological Tools. Eng, 5(4), 3259–3283. https://doi.org/10.3390/eng5040171

Rodríguez Elizalde, R. (2024b). Inspeção principal detalhada de construções civis antigas: exemplos de aplicação de drones para inspeção de construções Romanas. Brazilian Journal of Development, 10(9). https://doi.org/https://doi.org/10.34117/bjdv10n9-024

Rodríguez Elizalde, R. (2024c). Use of RPAS (DRONES) for Masonry Arch Bridges Inspection: Quality and Sustainable Work with Preventive Guarantee. Revista De Gestão Social E Ambiental, 18(12). https://doi.org/https://doi.org/10.24857/rgsa.v18n12-048

Rodríguez Elizalde, R. (2025). La inteligencia artificial ante la prevención de riesgos laborales: nuevos riesgos emergentes. In Inteligencia artificial y derecho: desafíos éticos, laborales y judiciales (pp. 164–173).

Rodríguez Sáenz de Galdeano, B. (2024). Daños derivados de la IA en el trabajo. Modelo regulador y responsabilidad civil. LABOS Revista de Derecho Del Trabajo y Protección Social, 5, 185–210. https://doi.org/https://doi.org/10.20318/labos.2024.9037

Rojas-Gualdron, D. F. (2022). La evaluación de tecnologías en salud basadas en inteligencia artificial ¿debería ser diferente? Revista CES Salud Pública y Epidemiología, 1(1), 53–58. https://doi.org/10.21615/cesspe.6572

Rojas, J. (2024). La inteligencia artificial (IA) y su relación con el área de la investigación. Revista de La Facultad de Farmacia, 65(2), 2–2. https://doi.org/10.53766/refa/2024.65.2.01

Ruiz, R. B., & Velásquez, J. D. (2023). Artificial intelligence at the service of the health of the future. Revista Medica Clinica Las Condes, 34(1), 84–91. https://doi.org/10.1016/j.rmclc.2022.12.001

Russell, S. (2017). Artificial intelligence: The future is superintelligent. Nature, 548(7669), 520–521. https://doi.org/10.1038/548520a

Sancho Escrivá, J. V., Fanjul Peyró, C., De la Iglesia Vayá, M., Montell, J. A., & Escartí Fabra, M. J. (2020). Aplicación de la Inteligencia Artificial con Procesamiento del Lenguaje Natural para textos de investigación cualitativa en la relación médico-paciente con enfermedad mental mediante el uso de tecnologías móviles. Revista de Comunicación y Salud, 10(1), 19–41. https://doi.org/10.35669/rcys.2020.10(1).19-41

Sanz-Valero, J., & Martínez-Aguirre, S. (2024). La inteligencia artificial y la salud laboral. Medicina y Seguridad Del Trabajo, 70(274), 6–9. https://doi.org/10.4321/s0465-546x2024000100001

Schor-Landman, G. (2022). El mundo ha cambiado. Alerta, Revista Científica Del Instituto Nacional de Salud, 5(1), 3–5. https://doi.org/10.5377/alerta.v5i1.13209

Seitech Solutions. (n.d.-a). Calidus, la suite de soluciones tecnológicas que ayuda a proteger tu empresa y a tus trabajadores. https://seitech.es/calidus-la-suite-de-soluciones-tecnologicas-que-ayuda-a-proteger-tu-empresa-y-a-tus-trabajadores/

Seitech Solutions (eFSO). (2023a). Piloto Calidus en Circet España. https://www.youtube.com/watch?v=eJNnVkRxLb0

Shimizu Corporation. (n.d.-a). Increasing Construction Site Productivity and Safety with Shimizu Smart Tunnel. The New Era of Tunnels. https://www.shimz.co.jp/en/topics/construction/item21

Shimizu Corporation. (n.d.-c). Next-Generation Tunnel Construction System: Shimizu Smart Tunnel. The New Era of Tunnels. https://www.shimz.co.jp/en/topics/civil/item04/index.html

Song, C., Ikei, H., & Miyazaki, Y. (2017). Sustained effects of a forest therapy program on the blood pressure of office workers. Urban Forestry and Urban Greening, 27, 246–252. https://doi.org/10.1016/j.ufug.2017.08.015

Soori, M., Arezoo, B., & Dastres, R. (2023). Machine learning and artificial intelligence in CNC machine tools, A review. Sustainable Manufacturing and Service Economics, 2, 100009. https://doi.org/10.1016/j.smse.2023.100009

Stefana, E., Di Gravio, G., & Patriarca, R. (2023). Adopting the Specific Operations Risk Assessment Methodology for Drone Inspections at Industrial Sites. 2023 7th International Conference on System Reliability and Safety, ICSRS 2023, 355–361. https://doi.org/10.1109/ICSRS59833.2023.10381275

Sugil, A. J. (2024). A Comprehensive Survey of Deep Learning and Its Applications in Advancing Artificial Intelligence. International Journal of Advanced Research in Computer Science, 15(2), 25–29. https://doi.org/10.26483/ijarcs.v15i2.7052

Taboada, M., Roldán, J. J., & Barrientos, A. (2022). Interfaz inteligente y adaptativa para personas con discapacidad. Actas de Las XXXVII Jornadas de Automática 7, 8 y 9 de Septiembre de 2016, Madrid, 483–490. https://doi.org/10.17979/spudc.9788497498081.0483

Terrones Rodríguez, A. L. (2018). Inteligencia artificial y ética de la responsabilidad. Cuestiones de Filosofía, 4(22), 141–170. https://doi.org/10.19053/01235095.v4.n22.2018.8311

The Factory School. (2025). IA y BIM: Casos reales de Inteligencia Artificial aplicada al diseño arquitectónico. The Factory School. https://thefactoryschool.com/blog/ia-y-bim-casos-reales/

Tirado Robles, C. (2022). El informe Future of Work 2035: For Everyone to Shine. Análisis jurídico de las propuestas japonesas sobre la implantación de los robots y de la IA en el mundo del trabajo y su posible adaptación al entorno europeo. Mirai. Estudios Japoneses, 6, 205–216. https://doi.org/10.5209/mira.80691

Tito Chura, V. F. (2023). Ciberseguridad Y Su Relación Con El Riesgo Cibernético, Ciudades Inteligentes, El Internet De Las Cosas (Iot) Y La Inteligencia Artificial. Revista Científica Seguridad y Desarrollo, 1(1). https://doi.org/10.58211/syd.v1i1.1

Toriumi, F. Y., Bittencourt, T. N., & Futai, M. M. (2023). UAV-based inspection of bridge and tunnel structures: an application review. Revista IBRACON de Estruturas e Materiais, 16(1). https://doi.org/10.1590/S1983-41952023000100003

Toro Lazo, A. (2024). El enfoque predictivo de la inteligencia artificial como herramienta para el mantenimiento de maquinaria industrial: una mirada desde la Industria 4.0. Entre Ciencia e Ingeniería, 18(35), 7–8. https://doi.org/10.31908/19098367.3130

Torres Abril, P. C., Jara Moya, S. D., Torres Valverde, L. D., & Arias Martínez, D. R. (2023). Revisión sistemática de las aplicaciones de vanguardia en el campo de la visión por computadora. Ciencia Digital, 7(4), 26–53. https://doi.org/10.33262/cienciadigital.v7i4.2710

Torres Cadillo, R. I. (2023). Los riesgos y los desafíos que enfrentan los trabajadores frente al uso de la inteligencia artificial en el trabajo. Revista de Derecho Procesal Del Trabajo, 6(7), 289–313. https://doi.org/10.47308/rdpt.v6i7.778

Tuya Gonzalez, P. J., Ramos Roman, I., & Dolado Cosin, J. J. (2011). Redes bayesianas en la ingenieria del software. Tecnicas Cuantitativas Para La Gestion En La Ingenieria Del Software, 205–222. https://doi.org/10.4272/978-84-9745-204-5.ch10

Uribe, I. (2024a). IA y Computer Vision: Prevención de riesgos en la construcción. Secmotic. https://secmotic.com/ia-computer-vision-construccion/#gref

Uribe, I. (2024b). IA y Computer Vision: Prevención de riesgos en la construcción. Secmotic.

Van Otterlo, M. (2009). Markov decision processes: concepts and algorithms. In Course on Learning and Reasoning (Issue May, pp. 1–23).

Vélez Aspiazu, E. E., Flores Beltrán, P. K., Flores Beltrán, C. M., Franco Zavala, A. Z., Chara Moreira, P. N., Rivera Castro, R. E., & Villamil Cavagnaro, C. J. (2024). Implementación de la inteligencia artificial en los proyectos arquitectónicos. LATAM Revista Latinoamericana de Ciencias Sociales y Humanidades, 5(5), 4871. https://doi.org/https://doi.org/10.56712/latam.v5i5.2949

Vuorinen, L., Aaltonen, K., Martinsuo, M., & Ahola, T. (2024). Finland. In Handbook on Public-Private Partnerships in International Infrastructure Development: A Critical Perspective (pp. 130–161). https://doi.org/10.4337/9781839102769.00012

Wagman, M. (2021). The Representation of Knowledge. Routledge Library Editions: Artificial Intelligence, 10–10, 57–70. https://doi.org/10.4324/9781351062909-6

Walton, P. (2018). Artificial intelligence and the limitations of information Information (Switzerland), 9(12). https://doi.org/10.3390/info9120332

Zaidi, S. F. A., Yang, J., Abbas, M. S., Hussain, R., Lee, D., & Park, C. (2024). Vision-Based Construction Safety Monitoring Utilizing Temporal Analysis to Reduce False Alarms. Buildings, 14(6). https://doi.org/10.3390/buildings14061878

Zamora-Cevallos, E., Zavala-Arteaga, L., Oviedo-Bayas, B., & Zúñiga-Delgado, M. S. (2024). Modelo predictivo para determinar patrones y tendencias que influyen en los accidentes de trabajo. Revista Metropolitana de Ciencias Aplicadas, 7(S2), 29–41. https://doi.org/10.62452/qrqpc986

Zapata Cortés, J. A. (2020). Inteligencia artificial para la toma de decisiones. Revista Perspectiva Empresarial, 7(2–1), 3–5. https://doi.org/10.16967/23898186.663

Zhang, R., Hao, G., Zhang, K., & Li, Z. (2023). Unmanned aerial vehicle navigation in underground structure inspection: A review. Geological Journal, 58(6), 2454–2472. https://doi.org/10.1002/gj.4763

www.ingramcontent.com/pod-product-compliance
Ingram Content Group UK Ltd.
Pitfield, Milton Keynes, MK11 3LW, UK
UKHW020657041026
15590UKWH00001B/317

9 798899 660276